CONCRETE LIQUID RETAINING STRUCTURES

Design, Specification and Construction

CONCRETE LIQUID RETAINING STRUCTURES

Design, Specification and Construction

J. KEITH GREEN

BSc, FICE, MIStructE, FIHE

and

PHILIP H. PERKINS

FASCE, FIMunE, FIArb, FIPHE, MIWES

APPLIED SCIENCE PUBLISHERS LTD
LONDON

APPLIED SCIENCE PUBLISHERS LTD
RIPPLE ROAD, BARKING, ESSEX, ENGLAND

British Library Cataloguing in Publication Data

Green, James Keith
Concrete liquid retaining structures
1. Concrete tanks—Design and construction
I. Title II. Perkins, Philip Harold
628.1'3 TA660.T34

ISBN 0-85334-856-1

WITH 20 TABLES AND 73 ILLUSTRATIONS

Printed in Great Britain by Galliard (Printers) Ltd, Great Yarmouth

CONTENTS

An expanded list of contents is included for Part 1 in order to show the general structure of the design calculations. It is also intended for use in the design office to provide quick and easy reference to the various design problems for which solutions are given. For Part 2, a comprehensive index is provided on pp. 351–3 .

Preface xi

Introduction xiii

PART 1. DESIGN

Notation 3

Chapter 1. Introductory notes on design 5
1.1. *Design* 5
1.2. *The influence of construction problems on design* 6
1.3. *Codes and handbooks* 6
1.4. *Elastic design* 6
1.5. *Permissible concrete stresses* 9
1.6. *Reinforcement stresses* 9
1.7. *Reinforcement spacing* 9
1.8. *Corner details* 10
1.9. *Bending moments in tank floors* 12
1.10. *Coefficients of thermal expansion* 13
1.11. *The design calculations* 13
Appendix: Prestressed and reinforced circular concrete tanks 15

Chapter 2. Design of a small reinforced concrete service water tank 18

DESIGN INFORMATION

2.1. *Design codes* 18
2.2. *Other design data* 18
2.3. *Loading conditions* 18
2.4. *Exposure conditions* 18
2.5. *Foundations* 19
2.6. *Material data and dimensions* 19

DESIGN NO. 1: LIMIT STATE DESIGN—CRACK CONTROL BY LIMITING STEEL STRESS

2.7. *Long walls* 20
2.8. *Short walls* 23
2.9. *Floor* 25
2.10. *Floor—short span* 26
2.11. *Floor—long span* 27
2.12. *Direct tension* 28
2.13. *Reinforcement for temperature and shrinkage* . 30
2.14. *Horizontal steel in short walls* 30
2.15. *Horizontal steel in long walls* 31
2.16. *Steel in floor—short span* 31
2.17. *Steel in floor—long span* 32
2.18. *Shear* 33
2.19. *Deflection* 34

DESIGN NO. 2: LIMIT STATE DESIGN—CRACK CONTROL BY CALCULATED CRACK WIDTH

2.20. *General approach* 35
2.21 *Long walls* 35
2.22. *Short walls* 36
2.23. *Floor—short span* 38
2.24. *Floor—long span* 38

DESIGN NO. 3: ALTERNATIVE METHOD OF DESIGN—BS 5337, CLAUSES 6, 9 AND 12

2.25. *General approach* 39
2.26. *Long walls* 40
2.27. *Short walls* 43
2.28. *Floor* 44
2.29. *Reinforcement arrangement* 45

Chapter 3. Design of an 'in ground' reinforced concrete swimming pool . . . 47

DESIGN INFORMATION

3.1. *Design codes* . . . 47
3.2. *Other design data* . . . 47
3.3. *Loading conditions* . . . 47
3.4. *Client's requirements* . . . 47
3.5. *Topography* . . . 48
3.6. *Borehole data* . . . 48
3.7. *Exposure conditions* . . . 49
3.8. *Design method* . . . 49
3.9. *Design principles* . . . 49

DESIGN CALCULATIONS

3.10. *Deflection* . . . 49
3.11. *Stability (full)* . . . 50
3.12. *Bearing pressures (full)* . . . 51
3.13. *Stability (empty)* . . . 53
3.14. *Bearing pressures (empty)* . . . 54
3.15(a). *Reinforcement in wall/water face (using Code Table 1)* . . . 54
3.15(b). *Reinforcement in wall/water face (alternative with crack width calculated)* . . . 55
3.15(c). *Reinforcement in wall/water face (alternative using tables in Handbook, Appendix 1)* . . . 58
3.16. *Reinforcement in wall/earth face* . . . 59
3.17. *Heel* . . . 59
3.18. *Wall corners* . . . 60
3.19. *Temperature and moisture effects* . . . 60
3.20. *Temperature and moisture effects—alternative approach* . . . 62

ALTERNATIVE DESIGN WITH MOVEMENT JOINTS

3.21. *Code Option 2 and proposed new Option 3* . . . 63

DESIGN OF POOL FLOOR

3.22. *Floor slab* . . . 66
3.23. *Horizontal tension due to water pressure* . . . 67

Chapter 4. Preliminary design of a circular prestressed concrete tank . . . 68

DESIGN INFORMATION

4.1. *Design codes* 68
4.2. *Other design data* 68
4.3. *Loading conditions* 68
4.4. *Client's requirements* 68
4.5. *Proposed design details* 69
4.6. *Design principles* 69
4.7. *Other parts of the structure* 70

DESIGN CALCULATIONS

4.8. *Ring stresses due to retained liquid* . . . 70
4.9. *Temperature stresses* 71
4.10. *Total ring stresses* 72
4.11. *Prestressing forces* 74
4.12. *Prestress losses* 75
4.13. *Prestressing steel* 76
4.14. *Maximum ring compression* 77
4.15. *Vertical bending moment due to contained liquid* 81
4.16 *Vertical bending due to temperature* . . 82
4.17. *Vertical reinforcement (outer face)* . . . 83
4.18. *Vertical bending due to earth pressure* . . 83
4.19. *Vertical bending due to prestressing* . . 83
4.20. *Vertical bending moment due to backfill and prestress (base)* 84
4.21. *Vertical bending moment at transfer (base)* . 84
4.22. *Vertical reinforcement (inner face)* . . . 85
4.23. *Vertical bending moment at mid-height* . . 85
4 24. *Vertical bending moment during prestressing* . 86
4.25. *Minimum reinforcement* 87
4.26. *Final notes* 87

PART 2. SPECIFICATION AND CONSTRUCTION

Chapter 5. Materials used in construction 93

Chapter 6. Factors controlling the durability of concrete . . 126

Chapter 7. Basic construction techniques 166

Chapter 8. Reinforced concrete liquid retaining structures constructed on or in the ground 202

Chapter 9. Prestressed concrete tanks, reinforced gunite tanks and water towers 259

Chapter 10. Quality control and testing 298

Chapter 11. Comments on British Standard BS 5337—*Code of Practice for the Structural Use of Concrete for Retaining Aqueous Liquids* 311

Appendix 1. Standards and Codes of Practice 317

Appendix 2. Conversion factors and constants 335

Appendix 3. Construction associations and research organisations 338

Appendix 4. Tolerances in construction 349

Index 351

PREFACE

The authors are conscious that this is the first completely new text book on the design and construction of concrete liquid retaining structures to be published in the United Kingdom for many years and is certainly the first to incorporate information which reflects the recommendations of the new Code of Practice on these structures (BS 5337:1976), together with its Handbook.

Both authors are practising engineers, and both were previously employed for many years with the Cement and Concrete Association to which organization they wish to record their debt for the experience that their work gave to them in the many applications of concrete.

They also acknowledge with gratitude the help they have received from many professional engineers and engineering firms. Special thanks are due to Lucy Perkins for her work in checking Part 2 of the manuscript as well as the page proofs and index, and to Richard Maling for his considerable assistance in the preparation and checking of the design examples in Part 1.

J. KEITH GREEN
PHILIP H. PERKINS

INTRODUCTION

This book is intended to provide practical information on the design, specification and construction of concrete liquid retaining structures. The fundamental principles of the structural design of these structures are illustrated by selected design examples based on the relevant Code of Practice, BS 5337 (formerly CP 2007).

While it has been written principally for practising engineers, the authors hope that it will prove useful to final year students in structural and civil engineering at Universities, Polytechnics, and Technical Colleges.

The opinions and recommendations in this book are those of the authors and are based on many years of experience. It is felt that they do not conflict with the Code, although they do not pretend to follow it in all cases. Some comments on the revised Code are given in Chapter 11, from which it can be seen that the authors have reservations about certain of the Code requirements. The Code, and other similar documents, are of great value, but they should not be treated as though they were perfect in all respects. They are compromise documents and embody the 'concensus' of opinion of a committee of experienced engineers. However, everyone who has attended meetings of professional men know how wide is the divergence of views expressed by the leading experts on a particular subject. British Standard Institution technical committees are no exception.

Limiting factors of many kinds have to be included in Codes and Standards and are intended for guidance, and there is seldom anything absolute about them, although in practice they are often rigidly enforced. These limiting factors are sometimes included without sufficient thought being given to the practical effect of their enforcement and their technical justification. Examples include the absorption of aggregates, the permitted concentration of chloride ions in reinforced and prestressed concrete and the use of admixtures in prestressed concrete pipes. All this adds up to the fact that there is no satisfactory substitute for the professional judgement of the engineer who is responsible for the design and supervision of the project.

In the course of their work the authors have come across cases where the designer has adopted an attitude towards the contractor which may be stated as follows:

> These are the requirements of the contract documents and you must follow them exactly, but even if there is no doubt that you have done so and the structure leaks, you are responsible and must put it right at your own expense.

It is acknowledged that the subject of contractual responsibility is a very difficult one, but such a contention is open to serious argument and the authors doubt if it would necessarily be accepted by the Courts or by an Arbitrator. For the satisfactory carrying through and completion of any major project, good site relations and co-operation between the Engineer and the Contractor are essential. Should serious bad feeling develop on a site, immediate steps should be taken to deal with it at top level.

The book has been divided into two parts. Part 1 deals with design from the point of view of the structural calculations needed to determine the dimensions of the concrete sections and the steel reinforcement required to safely withstand the applied loads.

Part 2 covers the specification of materials and workmanship and provides information on construction methods and problems encountered on site and how to overcome them.

In practice the two parts have to be considered as one, as it is impracticable and undesirable to draw a dividing line between design and construction.

The use of the two parts was decided for convenience of writing and presentation.

When considering design the engineer has to take into account how the structure will be built and the probable site conditions as far as these can be ascertained in advance. For example, a structure which will be built in an undeveloped country where there is a minimum of skilled labour, supervision and equipment is likely to be designed differently to a similar structure built in a highly industrialized country.

PART 1
Design

NOTATION

A_s	area of steel reinforcement
b	breadth (or width) of section
d	effective depth of tension reinforcement; diameter of tank
E_c	modulus of elasticity of concrete
E_s	modulus of elasticity of steel reinforcement
f_b	average bond strength between concrete and steel
f_c	compressive stress in concrete in bending
f_{ct}	direct tensile strength of the concrete; tensile stress in concrete
f_{cu}	characteristic cube strength of concrete at 28 days
f_{st}	stress in the tension reinforcement
f_t	ring tension per unit length
f_y	characteristic strength of the reinforcement
h	overall depth of member; depth of water
M	bending moment
M_d	design (service) moment of resistance
M_T	bending moment due to temperature
M_u	ultimate moment of resistance
p_{cb}	permissible compressive stress in concrete in bending
P_o	prestressing force in the tendon at the jacking end
P_x	prestressing force at distance x from the jack
q	distributed live load per unit length or per unit area
r	radius of tank
r_{ps}	radius of curvature of prestressing tendon
S_{max}	estimated maximum crack spacing
t	thickness of wall of tank
T_1	fall in temperature from hydration peak to ambient
T_2	seasonal fall in temperature
v	shear stress; shear force per unit length

V	total shear force
W_g	unit weight (e.g. 9·8 kN/m^3 for water) effective unit weight for soils (e.g. $k \times$ actual weight)
W_{max}	estimated maximum crack width
x	depth of the neutral axis
α	coefficient of thermal expansion of mature concrete
γ_f	partial safety factor for load
ε_{cs}	estimated shrinkage strain
ε_m	average strain at the level at which cracking is being considered, allowing for the stiffening effect of the concrete in the tension zone (see Appendix 1 of the Handbook)
ε_s	strain in the reinforcing steel
ε_{te}	estimated total thermal contraction strain after peak temperature due to heat of hydration
ε_1	strain at the level considered, ignoring the stiffening effect of the concrete in the tension zone
ρ_c	steel ratio based on gross concrete section
ρ_{crit}	critical steel ratio based on gross concrete section
ϕ	bar size; angle of internal friction
μ	coefficient of friction

Note: 1 N/mm^2 = 1 MN/m^2 = 1 MPa.

CHAPTER 1

INTRODUCTORY NOTES ON DESIGN

1.1. *Design*

Design for structures is the process by which it is planned that the structure will achieve its intended function in safety and at least overall cost.

For the structures with which this book is concerned, the function includes the ability to support the various possible combinations of loading together with the maintenance of an adequately watertight condition. In some circumstances other considerations may apply, such as thermal insulation, fire resistance and resistance to chemical attack. As soon as the structure is complete the influences of weather, ageing, use or misuse and other causes of deterioration begin to act. It is the designer's task to ensure the continued serviceability of the structure during its intended life.

Safety needs little explanation but the designer should always bear in mind that there are few structures to whose strength someone does not at some time entrust his life. The designer is seldom in a position to strike a proper balance between initial cost and maintenance cost. Clearly each may be reduced at the expense of the other. Unfortunately most clients have financial arrangements which separate these two elements of cost and pay for them from different funds. Nevertheless the origin of many costly maintenance problems lies in the original design and a substantial proportion of these could have been avoided or greatly reduced at little or no increase in initial cost. In water retaining structures, the inappropriate choice of type or location of movement joints and inadequate allowance for tolerances on reinforcement cover are responsible for the most frequent of such maintenance problems.

1.2. *The Influence of Construction Problems on Design*

It is therefore clear that design and construction are interrelated. The good designer cannot proceed without an understanding of the problems of construction; conversely the construction team will only achieve the designer's intentions if the site management has an adequate appreciation of the design. Indeed such considerations led to some discussion between the authors concerning the order in which the two parts of this book should be presented. Although chronological considerations prevailed, since design must always precede construction, the two parts are really inseparable. No designer should proceed without a careful study of Part 2.

1.3. *Codes and Handbooks*

In preparing the design sections of the book it has been assumed that the reader will have available copies of the following:

(a) BS 5337: 1976—*Code of Practice for the Structural Use of Concrete for Retaining Aqueous Liquids* (referred to in the text as 'the Code').
(b) *The Handbook to BS* 5337: 1976 (referred to in the text as 'the Handbook').
(c) CP 110: Parts 1 and 2: 1972—*The Structural Use of Concrete.*
(d) CP 114: Part 2: 1969—*The Structural Use of Reinforced Concrete in Buildings.*

A note of caution is appropriate at this point. Although the design examples in later chapters are based on the use of the appropriate Codes, these should not be regarded as embracing all the possible aspects of design. The Institution of Structural Engineers' concise report cited in the bibliography at the end of this chapter includes the following paragraph:

> The use of Codes is dangerous if their function is misunderstood and they are treated not as aids but as complete instructions. Under the impulse of economic competition, the Code is sometimes used as a 'check list'; satisfaction of its items is deemed automatically to ensure an adequate structure. This is an illusion.

1.4. *Elastic Design*

The limit state design method described in the Code implies the use of elastic design methods when checking the serviceability requirements. The

Code 'alternative method of design' specifically states (Clause 9.1) that design is to be based on elastic considerations.

Some years ago the use of design charts was common in elastic design of reinforced concrete rectangular beams and sections. The charts plotted the design parameter M/bd^2 against A_s/bd for various permissible steel stresses. In addition the corresponding neutral axis depth and concrete compressive stress could be read directly from the charts. Such charts enable the designer to obtain a 'feel' for the effect of changing one of the parameters. Accordingly the authors have derived the design chart shown in Fig. 1.1 which includes the steel and concrete stresses commonly required for the design of water retaining structures in accordance with BS 5337.

The method of use is largely self-evident but will be described for those not familiar with these charts. Having calculated the required service moment M this is divided by the selected b and d values. The chart is then entered on the vertical axis at the appropriate value of M/bd^2 and a horizontal line projected to intersect the appropriate permissible steel stress (f_{st}) line, The required A_s/bd value is read off on the bottom line immediately below the intersection point.

The corresponding concrete compressive stress is read off by interpolation between the two f_c lines closest to the point of intersection.

From the same point of intersection a vertical line may be dropped to intersect the x/d line allowing the value x/d to be read from the right-hand vertical scale. Where particular accuracy is required the actual value of A_s/bd for the particular reinforcement area chosen may be used and a vertical line drawn from this value to intersect the x/d line to give the particular value of x/d.

Example. A slab is to be reinforced to carry a service bending moment of 31·25 kN m with steel stress limited to 130 N/mm².

Try $d = 250$ mm and for a slab take $b = 1000$ mm:

$$\frac{M}{bd^2} = \frac{31{\cdot}25 \times 10^6}{1000 \times 250^2} = 0{\cdot}50$$

From which

$$\frac{A_s}{bd} = 0{\cdot}0042$$

and

$$\frac{x}{d} = 0{\cdot}30$$

Hence

$$A_s = 1000 \times 250 \times 0{\cdot}0042 = 1050\,\text{mm}^2/\text{m}$$

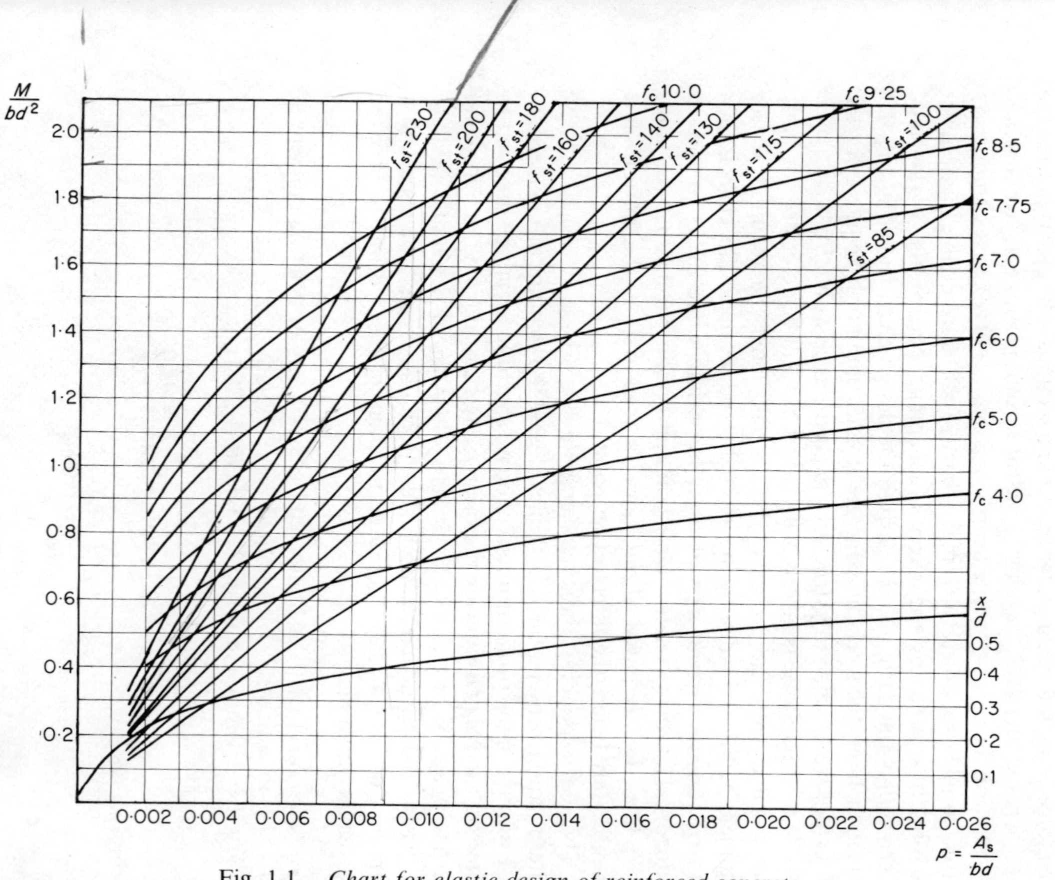

Fig. 1.1. *Chart for elastic design of reinforced concrete.*

12-mm bars at 100 c/c would suffice but suppose 16-mm bars at 160 c/c (1260 mm^2/m) are used to conform with other reinforcement. Then the actual A_s/bd

$$= \frac{1260}{1000 \times 250} = 0{\cdot}005$$

giving a slightly higher value of $x/d = 0{\cdot}32$. The difference in x (80 mm against 75 mm) is small in this instance but may be worth taking into account in crack width calculations.

1.5. *Permissible Concrete Stresses*

The Code does not directly limit the flexural concrete compressive stress but the Handbook indicates (section 8) that the value of $0{\cdot}45 f_{cu}$, given in Appendix C of the Code, should be taken. Hence for Grade 25 concrete $p_{cb} = 11{\cdot}3$ N/mm^2 and for Grade 30 concrete $p_{cb} = 13{\cdot}5$ N/mm^2. It will be noted that these values are somewhat higher than those permitted by CP 114. However, the crack width requirements for limit state design and the resistance to cracking calculations for the alternative method both generally have the effect of indirectly limiting the concrete stress to lower values.

1.6. *Reinforcement Stresses*

For limit state design when determining the ultimate limit state the Code follows the recommendations of CP 110. However, it must be noted that the characteristic strength of the reinforcement should not be taken as more than 425 N/mm^2 (Clause 8.1(b)). The higher values permitted in CP 110, Clause 3.1.4.3, cannot therefore be used. In flexural members the ultimate limit state is seldom critical and this is not therefore a severe limitation; in many designs it will be preferable to take a characteristic strength of 410 N/mm^2, thereby permitting the contractor a free choice between hot rolled high yield steel to BS 4449 or cold worked high yield steel to BS 4461.

Obviously if mild steel is used the characteristic strength is limited to 250 N/mm^2 in accordance with CP 110, Clause 3.1.4.3.

1.7. *Reinforcement Spacing*

The designer should always bear in mind the problems facing the steel fixer. It is convenient for setting out and checking the reinforcement if the bar spacings in the two faces of a slab are either equal or if the spacing on one face is a multiple of two or three times that on the other face.

Similar considerations apply where one reinforced slab meets another at right angles (e.g. at a floor/wall corner). Here the reinforcement in both faces of each slab must be selected to have a spacing which takes account of the intersection at the corner.

These requirements are such that it is often preferable to calculate the areas required and postpone selection of diameter and spacing until the reinforcement has been calculated for each face of all the intersecting slabs. Some over-provision of steel may be required if steel fixing problems are to be avoided.

In this respect it is unfortunate that the design tables for the limiting moments, given in Appendix 1 of the Handbook, consider bar spacings only in 50-mm increments between 100 mm and 300 mm. The text suggests that these spacings conform with the standard spacings recommended in report TRCS2 issued jointly by the Concrete Society and the Institution of Structural Engineers in 1968 (see the bibliography at the end of this chapter). However, the full range of standard bar spacings in TRCS2 is (mm) 50, 75, 100, 125, 150, 175, 200, 250, 300. In any case these recommendations were omitted entirely in the later revision of the report issued in 1970 by the same joint committee.

For many water retaining structures main reinforcing bars will be spaced in the range 100–200 mm. Direct use of the Handbook tables therefore results in, for example, a 50% increase in steel content if 150-mm spacing is just inadequate since the next tabulated spacing is 100 mm. Although not mentioned in the Handbook it is clear that some interpolation may be applied but this must be done with care. For example, referring to Table A1.1.10 of the Handbook, for 20-mm bars a moment of 611 kN m is given for 100-mm spacing and 428u kN m for 150-mm spacing. Not only is the difference too large for safe interpolation, but 611 kN m is based on the limiting crack width whereas the 'u' indicates that 428 kN m is based on the ultimate moment requirement. Interpolation between such tabulated values is therefore inadmissible.

1.8. *Corner Details*

Many water retaining structures include corner connections between two walls or between a wall and the floor. For corners where the main loading condition is such that the internal angle tends to close under load the

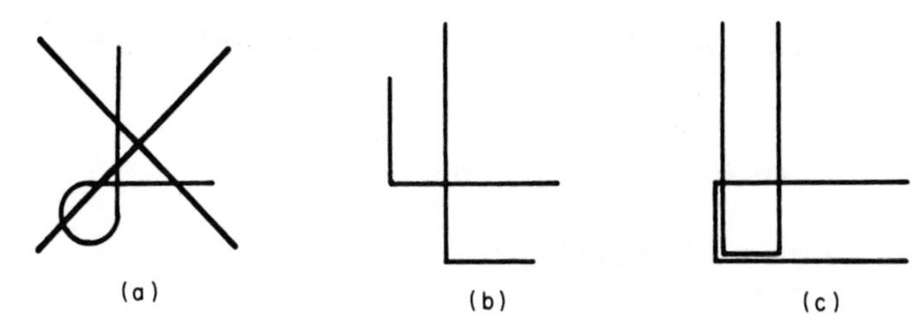

Fig. 1.2. *Corner bars.*

normal details shown in Fig. 1.2 (b) and (c) are satisfactory subject to the bars having adequate anchorage. Detail (a) should not be used.

However, in most water retaining structures the corners tend to open under the load from the internal water pressure. At the connection in a reservoir or swimming pool formed between a cantilever wall and its base the bending moment usually reaches its maximum value and is tending to open the corner. Under these conditions tests have shown that the bending moment transmitted by the details for closing corners shown in Fig. 1.2 (b) and (c) may be less than 25% of the flexural strength of the corresponding wall section. The intersecting 'hairpins' detail shown in Fig. 1.3 is recommended in these circumstances in the Concrete Society report '*Standard Reinforced Concrete Details*' published in 1973. Where it is practicable to do so the corner should be provided with a splay reinforced with diagonal bars.

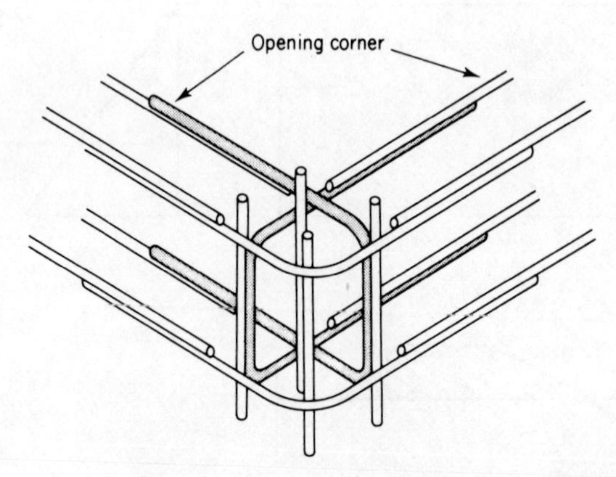

Fig. 1.3. *Arrangement of U-bars suitable for an opening corner.*

1.9. *Bending Moments in Tank Floors*

It is not immediately obvious that the maximum sagging moment in a tank floor, which is continuous with the walls, does not occur when the tank is full.

Consider the tank shown in Fig. 1.4 filled with liquid of density w kN/m^3 to a height of H m. The fixing moment M_f at the end of the floor is given by

$$M_f = -F \times \frac{H}{3} = -w\frac{H^3}{6}$$

Adding the maximum 'free' moment due to the contained liquid gives the maximum sagging moment:

$$M = -w\frac{H^3}{6} + \frac{wHL^2}{8}$$

and

$$\frac{\mathrm{d}M}{\mathrm{d}H} = \frac{-wH^2}{2} + \frac{wL^2}{8}$$

which equals zero when $H = L/2$. Hence the maximum sagging moment in the tank floor occurs when it is filled to a height equal to half the floor span.

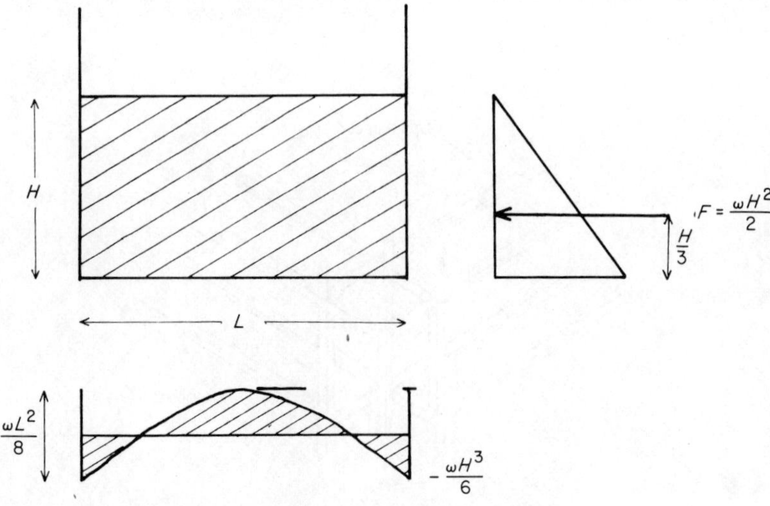

Fig. 1.4. *Bending moments in a tank floor.*

1.10. *Coefficients of Thermal Expansion*

The thermal expansion of concrete made with various aggregates is discussed in Part 2, Chapter 7. For normal design purposes it is usual to take a value of 12×10^{-6} per °C for the coefficient of thermal expansion of structural concrete.

In some circumstances it may be advantageous to use a lower value, for example where it is known that concrete containing limestone aggregate is to be incorporated. However, the lower values of thermal coefficient sometimes quoted should be used with caution. The thermal coefficient of cement paste is made up of two components, namely the true kinetic coefficient and swelling pressure. However, swelling does not occur when the specimen is dry or when it is saturated. It follows that at these two extremes the coefficient of thermal expansion is lower than when the paste is partially saturated. Neville (see the bibliography at the end of this chapter) quotes the values for thermal coefficients for 1:6 cement/aggregate concretes made with various aggregate types given in Table 1.1.

Table 1.1

Aggregate type	*Air-cured (64% RH),* 10^{-6} *per °C*	*Water-cured (saturated),* 10^{-6} *per °C*	*Wetted after air curing,* 10^{-6} *per °C*
Quartzite	13·1	12·2	11·7
Granite	9·5	8·6	7·7
Limestone	7·4	6·1	5·9

It is for the designer to decide whether in a particular application the concrete forming a water retaining structure will be permanently and completely saturated at all times. In other cases it would be advisable to use the higher values.

1.11. *The Design Calculations*

Chapters 2, 3 and 4 set out typical examples of calculations prepared generally in accordance with BS 5337.

Chapter 2 is intended for the student or engineer who is familiar with reinforced concrete design in accordance with the principles of CP 110 or CP 114. Each step is fully explained to an extent not normally included in a typical design office calculation. The simple rectangular tank is designed using each of the three main alternatives described in BS 5337. These are

(a) limit state design using allowable steel stresses for the serviceability limit state;
(b) limit state design using calculated crack widths;
(c) alternative method—modular ratio design using elastic theory.

In Chapter 3 the main calculations required for an 'in ground' swimming pool are illustrated. The calculations are set out to conform with normal design office practice but with comment interspersed where necessary to complete the explanations.

Chapter 4 illustrates the preliminary design calculations for a circular prestressed concrete tank. In a book of this type it is not practicable to include the complete design calculations for such a tank and for more detail the reader is referred to the specialist literature listed in the bibliography at the end of Chapter 4.

Bibliography

British Standards Institution. CP 114: Part 2: 1969—*Code of Practice for the Structural Use of Reinforced Concrete in Buildings*. London, 1969.
British Standards Institution. CP 110: Parts 1 and 2: 1972—*The Structural Use of Concrete*. London, 1972.
British Standards Institution. BS 5337: 1976—*Code of Practice for the Structural Use of Concrete for Retaining Aqueous Liquids* (formerly CP 2007). London, 1976.
British Standards Institution. BS 4449: 1978—*Specification for Hot Rolled Steel Bars for the Reinforcement of Concrete*. London, 1978.
British Standards Institution. BS 4461: 1978—*Specification for Cold Worked Steel Bars for the Reinforcement of Concrete*. London, 1978.
British Standards Institution. *Handbook to BS* 5337: 1976. London, 1979.
Concrete Society. *Standard Reinforced Concrete Details*. London, 1973.
Concrete Society/Institution of Structural Engineers. *The Detailing of Reinforced Concrete*. Technical Report TRCS2. London, 1968.
Concrete Society/Institution of Structural Engineers. *Standard Method of Detailing Reinforced Concrete*. Revision of Technical Report TRCS2. London, 1970.
Institution of Structural Engineers. *Aims of Structural Design*. London, 1969.
Neville, A. M. *Properties of Concrete*. Pitman Publishing. London, 1973.

Appendix: Prestressed and Reinforced Circular Concrete Tanks

The following tables give coefficients additional to those given in Appendix 3 of the Handbook to BS 5337 covering values of h^2/dt exceeding 16. (See also Chapter 4.) (The table numbers in each case correspond with those in the Handbook and the symbols are as given in the Handbook.)

Table A3.1a

h^2/dt	*Coefficients at point from top*				
	0·75 h	0·80 h	0·85 h	0·90 h	0·95 h
20	+0·716	+0·654	+0·520	+0·325	+0·115
24	+0·746	+0·702	+0·577	+0·372	+0·137
32	+0·782	+0·768	+0·663	+0·459	+0·182
40	+0·800	+0·805	+0·731	+0·530	+0·217
48	+0·791	+0·828	+0·785	+0·593	+0·254
56	+0·763	+0·838	+0·824	+0·636	+0·285

Table A3.2a

h^2/dt	*Coefficients at point from top*				
	0·75 h	0·80 h	0·85 h	0·90 h	0·95 h
20	+0·812	+0·817	+0·756	+0·603	+0·344
24	+0·816	+0·839	+0·793	+0·647	+0·377
32	+0·814	+0·861	+0·847	+0·721	+0·436
40	+0·802	+0·866	+0·880	+0·778	+0·483
48	+0·791	+0·864	+0·900	+0·820	+0·527
56	+0·781	+0·859	+0·911	+0·852	+0·563

Table A3.3a

h^2/dt	*Coefficients at point from top*				
	0·80 h	0·85 h	0·90 h	0·95 h	1·00 h
20	+0·0015	+0·0014	+0·0005	−0·0018	−0·0063
24	+0·0012	+0·0012	+0·0007	−0·0013	−0·0053
32	+0·0007	+0·0009	+0·0007	−0·0008	−0·0040
40	+0·0002	+0·0005	+0·0006	−0·0005	−0·0032
48	0·0000	+0·0001	+0·0006	−0·0003	−0·0026
56	0·0000	0·0000	+0·0004	−0·0001	−0·0023

Table A3.4a

h^2/dt	Coefficients at point from top				
	0·75 h	0·80 h	0·85 h	0·90 h	0·95 h
20	+0·949	+0·825	+0·629	+0·379	+0·128
24	+0·986	+0·879	+0·694	+0·430	+0·149
32	+1·026	+0·953	+0·788	+0·519	+0·189
40	+1·040	+0·996	+0·859	+0·591	+0·226
48	+1·043	+1·022	+0·911	+0·652	+0·262
56	+1·040	+1·035	+0·949	+0·705	+0·294

Table A3.5a

h^2/dt	Coefficients at point from top				
	0·75 h	0·80 h	0·85 h	0·90 h	0·95 h
20	+1·062	+1·017	+0·906	+0·703	+0·394
24	+1·066	+1·039	+0·943	+0·747	+0·427
32	+1·064	+1·061	+0·997	+0·821	+0·486
40	+1·052	+1·066	+1·030	+0·878	+0·533
48	+1·041	+1·064	+1·050	+0·920	+0·577
56	+1·021	+1·059	+1·061	+0·952	+0·613

Table A3.6a

h^2/dt	Coefficients at point from top				
	0·80 h	0·85 h	0·90 h	0·95 h	1·00 h
20	+0·001 5	+0·001 3	+0·000 2	−0·002 4	−0·007 3
24	+0·001 2	+0·001 2	+0·000 4	−0·001 8	−0·006 1
32	+0·000 8	+0·000 9	+0·000 6	−0·001 0	−0·004 6
40	+0·000 5	+0·000 7	+0·000 7	−0·000 5	−0·003 7
48	+0·000 4	+0·000 6	+0·000 6	−0·000 3	−0·003 1
56	+0·000 2	+0·000 4	+0·000 5	−0·000 1	−0·002 6

Table A3.7a

h^2/dt	*Coefficients at point from top*				
	0·75 h	0·80 h	0·85 h	0·90 h	0·95 h
20	+0·000 8	+0·001 4	+0·002 0	+0·002 4	+0·002 0
24	+0·000 5	+0·001 0	+0·001 5	+0·002 0	+0·001 7
32	0·000 0	+0·000 5	+0·000 9	+0·001 4	+0·001 3
40	0·000 0	+0·000 3	+0·000 6	+0·001 1	+0·001 1
48	0·000 0	+0·000 1	+0·000 4	+0·000 8	+0·001 0
56	0·000 0	0·000 0	+0·000 3	+0·000 7	+0·000 8

CHAPTER 2

DESIGN OF A SMALL REINFORCED CONCRETE SERVICE WATER TANK

DESIGN INFORMATION

2.1. *Design Codes*

BS 5337: 1976 (referred to as 'the Code').
CP 110: 1972.
CP 114: Part 2: 1969.

2.2. *Other Design Data*

Handbook to BS 5337*:* 1976 (referred to as 'the Handbook').

2.3. *Loading Conditions*

Water—density equivalent to 9·8 kN/m^3.
Reinforced concrete—density equivalent to 25 kN/m^3.
Wind—not applicable.
(Water temperature as ambient.)

2.4. *Exposure Conditions*

Class B of BS 5337, Clause 4.9.

Notes

(a) Although the top internal surfaces are subject to alternate wetting and drying, the flexural stresses at this level are very low. Hence the crack widths will automatically be small enough to provide for the Class A exposure implied for this position by the above clause—see also the Handbook, paragraph 4.9.

(b) The limit state designs which follow use concrete thicknesses less than 225 mm, and Class B exposure therefore applies both internally and externally. Thicknesses required for the 'alternative method' design would have permitted use of Class C exposure for the external faces but Class B has been used—see note under Section 2.29.

2.5. *Foundations*

Continuous rigid support around the perimeter of the tank is provided by a framed structure.

2.6. *Material Data and Dimensions*

Concrete—Grade 30.
Reinforcing steel—deformed high yield to BS 4449 or BS 4461.

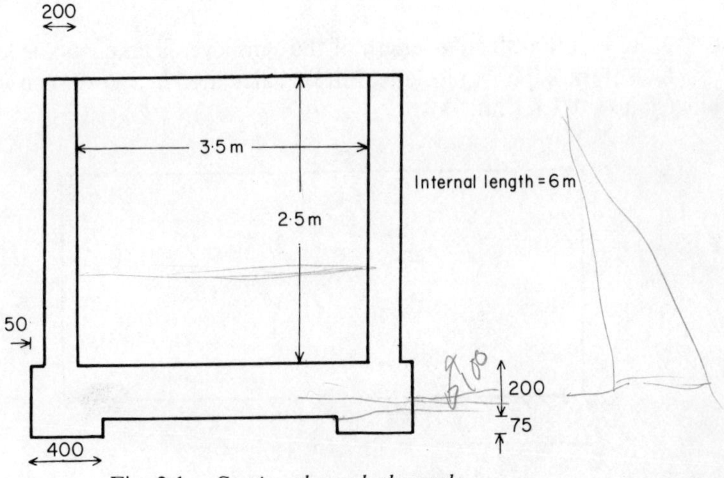

Fig. 2.1. *Section through the tank.*

The internal dimensions of the tank are 3·5 m × 6 m × 2·5 m high (Fig. 2.1).

DESIGN NO. 1: LIMIT STATE DESIGN—CRACK CONTROL BY LIMITING STEEL STRESS

2.7. Long Walls

Long walls are subject to primary bending in a vertical plane due to the effect of the triangularly distributed load (see Fig. 2.2). Support is effectively at the bottom edge only but provision will be made for secondary horizontal bending at the corners as for short wall design (Section 2.8 and see also Section 2.15).

Serviceability Limit State

Maximum water pressure

$$= 2{\cdot}5\, W_g = 2{\cdot}5 \times 9{\cdot}8 = 24{\cdot}5\,\text{kN/m}^2$$

and the bending moment on a section of wall 1 m wide

$$= 24{\cdot}5 \times \frac{2{\cdot}5}{2} \times \left(\frac{2{\cdot}5}{3} + 0{\cdot}1\right) = 28{\cdot}6\,\text{kN m}$$

Note that the effective length of the cantilever is taken as the length to the face of the support plus an addition calculated in accordance with CP 110, Clauses 3.3.1.1 and 3.4.1.

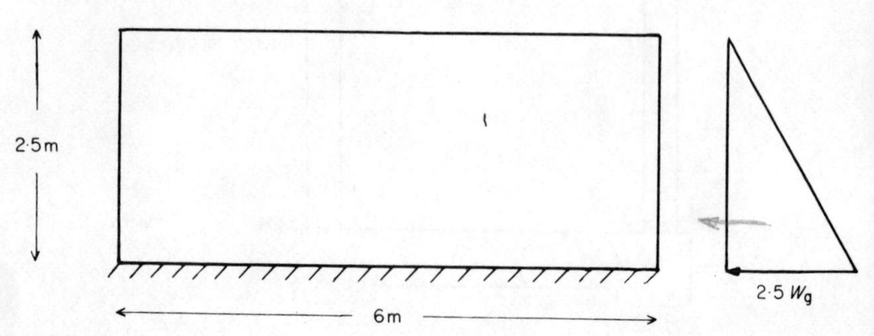

Fig. 2.2. *Triangularly distributed load on long walls.*

Try $h = 200$ mm and $d = 140$ mm (note that distribution reinforcement is to be placed outside the main reinforcement in accordance with Clause 4.11 of the Code—allow 40-mm cover to distribution steel and, say, 60-mm to centre of main bars):

$$\frac{M}{bd^2} = \frac{28\cdot6 \times 10^6}{1000 \times 140^2} = 1\cdot46$$

where b is taken as the breadth of a 1·0-m-wide strip.

From the design chart for elastic design (Fig. 1.1)

$$p = \frac{A_s}{bd} = 0\cdot0133$$

for $f_{st} = 130\ \text{N/mm}^2$ (and $f_c = 7\cdot5\ \text{N/mm}^2$, i.e. less than $0\cdot45 f_{cu}$).

$$A_s = 0\cdot0133 \times 1000 \times 140 = 1862\ \text{mm}^2/\text{m}$$

Y20 @ 170 c/c

Check Ultimate Limit State

Taking γ_f as 1·6

$$M_u = 1\cdot6 \times 28\cdot6 = 45\cdot8\ \text{kN m}$$

$$\frac{M_u}{bd^2} = \frac{45\cdot8 \times 10^6}{1000 \times 140^2} = 2\cdot34$$

From CP 110: Part 2, Design Chart 2, since $f_y = 410\ \text{N/mm}^2$ for hot rolled high yield steel

$$\frac{100A_s}{bd} = 0\cdot73$$

Therefore

$$A_s = 0\cdot73 \times 1000 \times \frac{140}{100} = 1022\ \text{mm}^2/\text{m}$$

Satisfactory

which is less than that provided for the serviceability condition. The required reinforcement is now entered in the output margin against the value of A_s calculated for serviceability.

Now consider curtailment of some of this vertical steel. Table 7.1 of the Handbook gives a lap length of 53ϕ, i.e. 1060 mm for a 20-mm bar.

However, this may be reduced in the ratio of reinforcement area required for ultimate limit state to the area actually provided, i.e.

$$1060 \times \frac{1022}{1862} = 582\,\text{mm}$$

It will be convenient to terminate the wall starter bars at, say, 600 mm above the 'kicker'. This is normally at least 75 mm above the floor surface.

Serviceability Limit State

For the serviceability limit state the water pressure on the wall at 0·675 m above the floor

$$= 1{\cdot}825 \times 9{\cdot}81 = 17{\cdot}9\,\text{kN/m}^2$$

and

$$M = 17{\cdot}9 \times \frac{1{\cdot}825}{2} \times \frac{1{\cdot}825}{3} = 9{\cdot}9\,\text{kN m}$$

$$\frac{M}{bd^2} = \frac{9{\cdot}9 \times 10^6}{1000 \times 140^2} = 0{\cdot}51$$

for which Fig. 1.1 gives

$$\frac{A_s}{bd} = 0{\cdot}0043$$

and

$$A_s = 0{\cdot}0043 \times 1000 \times 140 = 602\,\text{mm}^2/\text{m}$$

*Y*12 @ 170 *c*/*c*

Check Ultimate Limit State

$$M_u = 1{\cdot}6 \times 9{\cdot}9 = 15{\cdot}9\,\text{kN m}$$

$$\frac{M_u}{bd^2} = \frac{15{\cdot}9 \times 10^6}{1000 \times 140^2} = 0{\cdot}81$$

for which

$$\frac{100\,A_s}{bd} = 0{\cdot}25$$

and

$$A_s = 0{\cdot}25 \times 1000 \times \frac{140}{100} = 350\,\text{mm}^2/\text{m}$$

which is less than that provided for serviceability.

Satisfactory

2.8. *Short Walls*

Short walls are subject to bending vertically and horizontally, due to the effect of the triangularly distributed load, and support is at two vertical edges and the bottom edge (Fig. 2.3). Increase height and breadth by 100 mm to give effective spans.

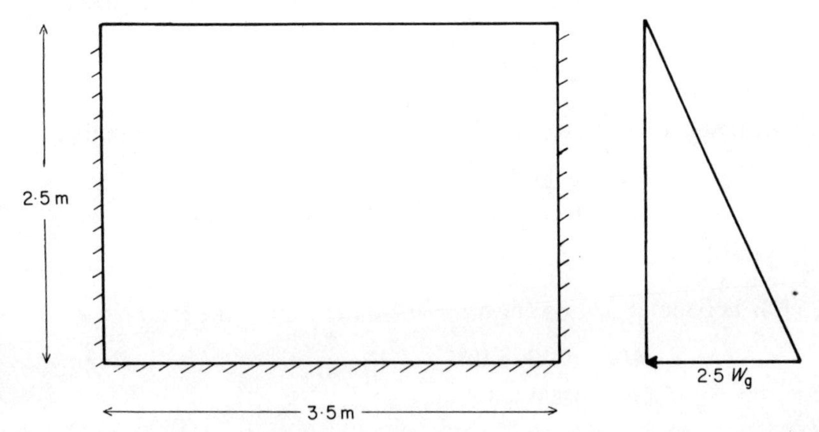

Fig. 2.3. *Triangularly distributed load on short walls.*

Serviceability Limit State
Maximum water pressure

$$= 2{\cdot}5\,W_g = 2{\cdot}5 \times 9{\cdot}8 = 24{\cdot}5\,\text{kN/m}^2$$

Span ratio l_x/l_z

$$= \frac{3{\cdot}6}{2{\cdot}6} = 1{\cdot}4$$

where l_x is the effective horizontal span on the short wall and l_z is the effective vertical span.

From the Handbook, Table A4.1, Case 1, it is found that height

$$\lambda_2 = \lambda_3 = 0{\cdot}55$$

Hence maximum horizontal bending moment occurs at a height of

$$0{\cdot}55 \times 2{\cdot}5 = 1{\cdot}375\,\text{m}$$

Maximum positive horizontal service bending moment (corners)

$$= 24{\cdot}5 \times 3{\cdot}6^2 \times 0{\cdot}018 = 5{\cdot}72\,\text{kN m/m}$$

Maximum negative horizontal service bending moment (centre)

$$= 24{\cdot}5 \times 3{\cdot}6^2 \times 0{\cdot}03 = 9{\cdot}53\,\text{kN m/m}$$

Maximum vertical service bending moment

$$= 24{\cdot}5 \times 2{\cdot}6^2 \times 0{\cdot}058 = 9{\cdot}60\,\text{kN m/m}$$

For horizontal reinforcement on the inner face near the corners

$$\frac{M}{bd^2} = \frac{5{\cdot}72 \times 10^6}{1000 \times 155^2} = 0{\cdot}24$$

(Note that d now refers to the outer layer of reinforcement.) From Fig. 1.1

$$p = 0{\cdot}0020$$
$$A_s = 0{\cdot}0020 \times 1000 \times 155 = 310\,\text{mm}^2/\text{m}$$

See Section 2.14

For horizontal reinforcement on the outer face at the centre

$$\frac{M}{bd^2} = \frac{9{\cdot}53 \times 10^6}{1000 \times 155^2} = 0{\cdot}40$$

$$p = 0{\cdot}0034$$
$$A_s = 0{\cdot}0034 \times 1000 \times 155 = 527\,\text{mm}^2/\text{m}$$

See Section 2.14

For vertical reinforcement on the inner face

$$\frac{M}{bd^2} = \frac{9{\cdot}60 \times 10^6}{1000 \times 140^2} = 0{\cdot}49$$

$$p = 0{\cdot}0041$$

$$A_s = 0{\cdot}0041 \times 1000 \times 140 = 574\,\text{mm}^2/\text{m}$$

*Y*12 @ 180 *c/c (slightly increased to suit bar spacing in floor—Section* 2.17)

Check Ultimate Limit State

For horizontal bending moment on the inner face

$$\frac{M_u}{bd^2} = 1{\cdot}6 \times 0{\cdot}24 = 0{\cdot}38$$

and CP 110: Part 2, Chart 2 gives

$$\frac{100A_s}{bd} = 0{\cdot}13$$

Hence

$$A_s = 0{\cdot}13 \times 1000 \times \frac{155}{100} = 201\ \text{mm}^2/\text{m}$$

Satisfactory

and the ultimate limit state requirements for horizontal reinforcement on the outer face and for vertical reinforcement on the inner face are similarly less than those for the serviceability limit state previously calculated. The required steel may now be entered in the output margin against the serviceability calculation for vertical steel. Horizontal reinforcement selection is deferred pending consideration of direct tension in Section 2.14.

2.9. *Floor*

The maximum sagging bending moment in the floor slab does not necessarily occur when the tank is full due to the reduced fixity at the edges in a partially full tank. Maximum sagging moment occurs when the water height is equal to half the effective span (see Chapter 1). Effective spans are 6·2 m and 3·7 m, and referring to CP 110, Table 12

$$\frac{l_y}{l_x} = \frac{6{\cdot}2}{3{\cdot}7} = 1{\cdot}68$$

where l_x is the length of the shorter side and l_y is the length of the longer side, so that the moment coefficients become

$$\alpha_{sx} = 0{\cdot}11$$

and

$$\alpha_{sy} = 0{\cdot}04$$

Whilst these figures from Table 12 apply specifically to ultimate loads they are based on an elastic analysis and may therefore be used both for the ultimate and serviceability limit states.

2.10. *Floor—Short Span*

Serviceability Limit State

Sagging moment in the short span is a maximum when water depth

$$= \frac{3{\cdot}7}{2} = 1{\cdot}85\,\text{m}$$

Pressure on the wall at the bottom of the tank is now $1{\cdot}85\,W_g$

$$= 1{\cdot}85 \times 9{\cdot}81$$
$$= 18{\cdot}2\,\text{kN/m}^2$$

and the bending moment at the wall base

$$= 18{\cdot}2 \times \frac{1{\cdot}85}{2} \times \frac{1{\cdot}85}{3}$$
$$= 10{\cdot}4\,\text{kN m/m}$$

Self weight of the floor

$$= 25 \times 0{\cdot}200$$
$$= 5\,\text{kN/m}^2$$

The maximum 'free' bending moment in the short span direction therefore

$$= 0{\cdot}11 \times (5 + 18{\cdot}2) \times 3{\cdot}7^2$$
$$= 34{\cdot}9\,\text{kN m/m}$$

and the total moment at mid-span

$$= 34{\cdot}9 - 10{\cdot}4 = 24{\cdot}5\,\text{kN m/m}$$

$$\frac{M}{bd^2} = \frac{24{\cdot}5 \times 10^6}{1000 \times 140^2} = 1{\cdot}25$$

From Fig. 1.1

$$p = \frac{A_s}{bd} = 0{\cdot}0113$$

$$A_s = 0{\cdot}0113 \times 1000 \times 140 = 1582\,\text{mm}^2/\text{m}$$

See Section 2.16

Check Ultimate Limit State

$$M_u = \{0{\cdot}11 \times [(1{\cdot}4 \times 5) + (1{\cdot}6 \times 18{\cdot}2)] \times 3{\cdot}7^2\} - (1{\cdot}6 \times 10{\cdot}4)$$
$$= 37{\cdot}8\,\text{kN m}$$

$$\frac{M_u}{bd^2} = \frac{37{\cdot}8 \times 10^6}{1000 \times 140^2} = 1{\cdot}93$$

and from CP 110: Part 2, Design Chart 2

$$\frac{100A_s}{bd} = 0{\cdot}58$$

$$A_s = 0{\cdot}58 \times 1000 \times \frac{140}{100} = 812\,\text{mm}^2/\text{m}$$

which is less than that provided for the serviceability condition.

Satisfactory

2.11. *Floor—Long Span*

Serviceability Limit State

Repeating as for the short span the maximum bending moment would occur when the water depth exceeds the tank depth—hence the full tank is taken. The bending moment at the wall base is therefore 9·6 kN m/m (see calculations for short wall, Section 2.8).

Maximum 'free' moment in the long span direction

$$= 0{\cdot}04 \times (5 + 24{\cdot}5) \times 3{\cdot}7^2$$
$$= 16{\cdot}2\,\text{kN m/m}$$

and the total moment at mid-span

$$= 16{\cdot}2 - 9{\cdot}6 = 6{\cdot}6\,\text{kN m/m}$$

Hence

$$\frac{M}{bd^2} = \frac{6{\cdot}6 \times 10^6}{1000 \times 155^2} = 0{\cdot}28$$

(Note that d now relates to the outer layer of reinforcement.) Figure 1·1 gives

$$p = 0{\cdot}0024$$

Therefore

$$A_s = 0{\cdot}0024 \times 1000 \times 155 = 372\,\text{mm}^2/\text{m}$$

See Section 2.17

Check Ultimate Limit State

$$M_u = \{0{\cdot}04 \times [(1{\cdot}4 \times 5) + (1{\cdot}6 \times 24{\cdot}5)] \times 3{\cdot}7^2\} - (1{\cdot}6 \times 9{\cdot}6)$$
$$= 9{\cdot}9\,\text{kN m}$$

and

$$\frac{M_u}{bd^2} = \frac{9{\cdot}9 + 10^6}{1000 \times 155^2} = 0{\cdot}41$$

CP 110: Part 2, Design Chart 2 gives

$$\frac{100A_s}{bd} = 0{\cdot}12$$

$$A_s = 0{\cdot}12 \times 1000 \times \frac{155}{100} = 186\,\text{mm}^2/\text{m}$$

which is less than that required for serviceability.

Satisfactory

2.12. *Direct Tension*

Now consider the direct tension induced in the floor and the short walls by the water pressure on the adjacent long walls. Referring to Fig. 2.1, the section properties given in Table 2.1 are derived.

$$\bar{y} = \frac{1{\cdot}754}{2{\cdot}360} = 0{\cdot}743\,\text{m}$$

Table 2.1

Part	*Dimensions, m*	*Area* (A), m^2	*Height to* *CG*(y), *m*	Ay	*Moment of inertia* (I), m^4	$A(\bar{y} - y)^2$
Walls	2(0·2 × 2·5)	1·000	1·525	1·525	0·5208	0·6115
Floor	6·5 × 0·20	1·300	0·175	0·227	0·0043	0·4194
Floor	2(0·4 × 0·075)	0·060	0·037	0·002	0·0003	0·0299
		2·360		1·754	0·5254	1·0608

Hence the moment of inertia of the whole section

$$= 0{\cdot}5254 + 1{\cdot}0608 = 1{\cdot}586\,\mathrm{m}^4$$

Total force on short walls and floor due to water pressure (Fig. 2.2)

$$= \frac{24{\cdot}5}{2} \times 6{\cdot}0 \times 2{\cdot}5$$

$$= 184\,\mathrm{kN}$$

and this acts at one-third height, i.e.

$$\frac{2{\cdot}500}{3} + 0{\cdot}275 = 1{\cdot}108\,\mathrm{m}$$

above the base and

$$1{\cdot}108 - 0{\cdot}743 = 0{\cdot}365\,\mathrm{m}$$

above the CG of the section. Hence tensile stress at top of short walls

$$= \frac{184 \times 10^3}{2{\cdot}36 \times 10^6} + \frac{184 \times 0{\cdot}365 \times 10^6}{Z_t}$$

where the section modulus referring to the top of the wall Z_t

$$= \frac{1{\cdot}586 \times 10^{12}}{(2{\cdot}775 - 0{\cdot}743) \times 10^3} = 781 \times 10^6$$

and tensile stress

$$= 0{\cdot}078 + 0{\cdot}086 \simeq 0{\cdot}16\,\mathrm{N/mm}^2$$

Similarly at the bottom of the wall the tensile stress

$$= 0{\cdot}078 - 0{\cdot}031 \simeq 0{\cdot}05\,\mathrm{N/mm}^2$$

Hence the tensile force to be resisted in each wall is

$$\frac{(0{\cdot}16 + 0{\cdot}05)}{2} \times 200 \times 2500 = 53\ \mathrm{kN}$$

or

$$\frac{53}{2{\cdot}5} = 21\ \mathrm{kN/m}$$

Additional area of horizontal steel acting at same stress as reinforcement provided for bending stresses (see Section 4.3 of the Handbook)

$$= \frac{21\,000}{130} = 162\,\mathrm{mm}^2/\mathrm{m}$$

See Section 2.14

Similarly in the floor the tensile force

$$= 184 - (2 \times 53) = 78\,\text{kN}$$

or

$$\frac{78}{6{\cdot}5} = 12\,\text{kN/m}$$

and additional steel

$$= \frac{12\,000}{130} = 92\,\text{mm}^2/\text{m}$$

See Section 2.17

2.13. *Reinforcement for Temperature and Shrinkage*

Now consider the reinforcement required to control cracking due to temperature and shrinkage (see Section 4.3 of the Handbook). For this small tank the provision of reinforcement for these purposes will be to Clause 4.11.2 of the Code, i.e. in each face of each wall and the floor

$$A_s = \frac{0{\cdot}15}{100} \times 200 \times 1000 = 300\,\text{mm}^2/\text{m}$$

2.14. *Horizontal Steel in Short Walls*

The vertical steel has already been determined and is not affected by direct tension. The horizontal steel previously determined from flexural considerations (Section 2.8) must, however, now be supplemented by steel to carry the direct tension. This extra steel is *not* additional to the reinforcement provided to control cracking due to temperature and moisture changes (Section 4.3 of the Handbook, final paragraph). It is therefore advantageous to distribute the additional steel required to meet the direct tension requirements so that it also meets the nominal reinforcement requirements.

Referring back to the flexural steel, 310 mm²/m was required on the inner face and 527 mm²/m on the outer face. Direct tension requires 162 mm²/m (Section 2.12).

If the tension steel is equally divided between the two faces, steel required on the inner face

$$= 391\,\text{mm}^2/\text{m}$$

*Y*10 @ 180 *c*/*c*

and steel required on the outer face

$$= 608\,\text{mm}^2/\text{m}$$

Y12 @ 180 c/c

and this is more than the minimum of 300 mm²/m required to control cracking due to temperature and shrinkage.

2.15. *Horizontal Steel in Long Walls*

Although these walls are designed for bending in one direction there will be local moments induced near the corners. For this reason the horizontal steel in the short wall will be carried round the corner to lap with the horizontal steel in the long walls. It will therefore simplify detailing and steel fixing if the spacing of horizontal bars in the long wall is either equal to, or a multiple of, that in the short walls, e.g. either 180 mm or 90 mm.

The steel required to resist direct tension in this direction has not been separately calculated and for this tank will be taken as equal to that for the short walls. Since the tensile force is reduced in the ratio

$$\frac{3{\cdot}5}{6{\cdot}0} = 0{\cdot}58$$

whilst the concrete area is only reduced in the ratio

$$\frac{(2{\cdot}36 - 0{\cdot}50)}{2{\cdot}36} = 0{\cdot}79$$

this assumption is conservative. Thus, the requirement for direct tension is a total of 162 mm²/m as before (Section 2.12).

In this case, this requirement is exceeded by the nominal steel requirement of 300 mm²/m in each face (Section 2.13).

Y10 @ 180 c/c both faces

2.16. *Steel in Floor—Short Span*

Previous calculations (Sections 2.10 and 2.7) showed that for flexural requirements

$$A_s\text{ (bottom at mid-span)} = 1582\,\text{mm}^2/\text{m}$$
$$A_s\text{ (top at supports)} = 1862\,\text{mm}^2/\text{m}$$

and for direct tension (Section 2.12)

$$A_s = 92\,\text{mm}^2/\text{m}$$

Y20 @ 170 c/c top and bottom (to conform with spacing of vertical steel in long walls)

However, for this short span, consideration of anchorage requirements, laps and ease of steel fixing indicates that there is no economic benefit to be gained by curtailing either the top or bottom steel. If there is no curtailment then 50% of the bottom steel is in excess of the minimum requirement at the supports and 50% of the top steel is similarly in excess at mid-span. Hence the requirement for the small direct tension is adequately met by the flexural steel if this is carried through the span.

2.17. *Steel in Floor—Long Span*

From previous calculations (Sections 2.11 and 2.8) for flexure

$$A_s \text{ (bottom at mid-span)} = 372\,\text{mm}^2/\text{m}$$
$$A_s \text{ (top at supports)} = 574\,\text{mm}^2/\text{m}$$

(for which Y12 at 180 c/c was provided in the walls) and for direct tension (Section 2.12)

$$A_s = 92\,\text{mm}^2/\text{m}$$

Additional steel will be provided for the direct tension in this case. Hence total steel required will be

$$\text{Bottom face} = 372 + \frac{92}{2} = 418\,\text{mm}^2/\text{m}$$

Bottom face Y10 @ 180 c/c

$$\text{Top face} = 574 + \frac{92}{2} = 620\,\text{mm}^2/\text{m}$$

Top face Y12 @ 180 c/c

2.18. *Shear*

Check Shear Due to Ultimate Loads

For the walls at the base

$$V = 24{\cdot}5 \times \frac{2{\cdot}5}{2} \times 1{\cdot}6 = 49{\cdot}0\,\text{kN/m}$$

For the floor slab (ignoring two-way spanning and taking as simply supported in the short span direction)

$$V = \left(24{\cdot}5 \times \frac{3{\cdot}5}{2} \times 1{\cdot}6\right) + \left(5 \times \frac{3{\cdot}5}{2} \times 1{\cdot}4\right) = 80{\cdot}9\,\text{kN/m}$$

For this value (CP 110, Clause 3.4.5.1)

$$v = \frac{80{\cdot}9 \times 10^3}{1000 \times 140} = 0{\cdot}58\,\text{N/mm}^2$$

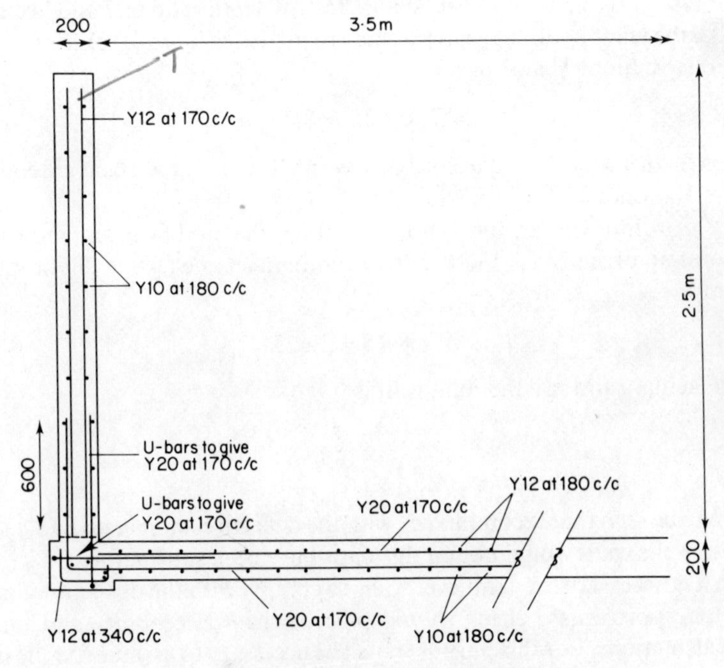

Fig. 2.4. *Reinforcement for Design No.* 1.

From CP110, Tables 5 and 14, shear reinforcement is not required for values of shear stress

$$\not> 1{\cdot}20 \times 0{\cdot}73 \left(\text{for} \frac{100A_s}{bd} = 1{\cdot}13 \right)$$

$$= 0{\cdot}88 \text{ N/mm}^2$$

Hence no shear reinforcement is required in the walls or floor. This is normally the case for slabs in water retaining structures. The reinforcement details can now be drawn as in Fig. 2.4.

2.19. *Deflection*

Deflection Limit State

Section 5.3.1 of the Handbook indicates that the horizontal deflection at the top of the walls need not be considered for a tank of this type. As an illustration, however, this will now be checked.

CP 110, Table 8, gives a basic span/effective depth ratio of 7 and Section 5.3.1 of the Handbook suggests this may be increased by 25 % for triangular load distribution. Hence basic ratio

$$= 7 \times 1{\cdot}25 = 8{\cdot}75$$

This is further modified in accordance with CP 110, Table 10, as extended by the Handbook, Table 5.3.1.

For $f_{st} = 130 \text{ N/mm}^2$ and $100A_s/bd = 1{\cdot}33$, the modification factor for tension reinforcement is 1·4 and the maximum span/effective depth ratio becomes

$$1{\cdot}4 \times 8{\cdot}75 = 12{\cdot}25$$

The actual ratio for the long wall is

$$\frac{2{\cdot}6}{0{\cdot}14} = 18{\cdot}6$$

It may be seen that if compliance with the deflection limit state had been necessary the walls would not on this basis have been satisfactory. In cases where it is necessary for cantilever walls to comply with the deflection limit state, an approximate check should precede the serviceability and limit state calculations. For this purpose it will be necessary to assume a value of, say, 1·0 for $100A_s/bd$.

For the floor the basic ratio from CP 110, Table 8, is 26. The modification factor for tension reinforcement $(100A_s/bd) = 1{\cdot}13$ from the Handbook, Table 5.3.1, is 1·49 and the maximum ratio becomes

$$1{\cdot}49 \times 26 = 38{\cdot}7$$

The actual ratio is

$$\frac{3{\cdot}7}{0{\cdot}14} = 26{\cdot}4$$

which is satisfactory.

DESIGN NO. 2: LIMIT STATE DESIGN—CRACK CONTROL BY CALCULATED CRACK WIDTH

2.20. *General Approach*

In the preceding calculations the crack width under service loads was 'deemed to be satisfactory' by limiting the steel stresses to those in Table 1 of the Code.

The alternative method of calculating crack widths is described in some detail in the Code, Clause 8.2 and Appendices B and C. Unfortunately the direct use of the formulae given in Appendix C for calculating crack widths due to flexure results in a tedious calculation which renders these unsuitable for normal design office application. For this reason Appendix 1 of the Handbook gives the limiting moments for a series of concrete section depths used in combination with tabulated reinforcement size and spacing. The tables have been prepared to give the moments corresponding to the two values of maximum crack width recommended in the Code, namely 0·1 mm for Class A exposure conditions and 0·2 mm for Class B exposure conditions.

These design tables will now be applied to the small tank previously designed by the 'deemed to satisfy' stress method.

2.21. *Long Walls*

Referring to the cantilever walls, the maximum service moment is 28·6 kN m/m calculated as before (Section 2.7).

Table A1.2.1 of the Handbook applies to a concrete section 200 mm thick with crack width limited to 0·2 mm. It is seen that the steel required is Y20 at 200 c/c (giving a limiting moment of 28·7 kN m/m at a stress of 155 N/mm^2).

Y20 @ 200 c/c at base

Similarly at the curtailment point used previously, the required moment capacity (9·9 kN m/m, Section 2.7) is given by Y12 at 300 c/c (giving a limiting moment of 11·8 kN m/m at a stress of 239 N/mm^2).

Use Y12 @ 200 c/c ab. kicker (see note (c) below)

It will be noted that this latter figure for moment is followed in the table by 'u' indicating that the ultimate moment is the limiting factor rather than the service moment for this particular section.

Several comparisons may now be made with the previous calculations for Design No. 1 which used 'deemed to satisfy' stress limits.

(a) Y20 at 200 c/c at the base of the long walls compares with Y20 at 170 c/c (Design No. 1);

(b) Y12 at 300 c/c above the curtailment point compares with Y12 at 170 c/c (Design No. 1);

(c) Calculated steel stress (Design No. 1) was 130 N/mm^2 against the tabulated 155 N/mm^2 and 239 N/mm^2 for (a) and (b) above. The latter figure is high but should be reduced proportionately to allow for actual service moment/tabulated service moment to give a corrected value of 200 N/mm. A note of caution against using the highest steel stresses permitted by the Code and included in the tables is given in Section A1.4(4) of the Handbook. The authors strongly endorse this comment. In this case, a reduction of the spacing for the Y12 bars to 200 mm c/c, to conform with the spacing for the Y20 bars at the wall base, is appropriate for construction reasons and also reduces the steel stress to 133 N/mm^2.

Note that when these tables are used, it is not necessary to carry out a separate check on the ultimate moment since this has already been allowed for in preparing the tables.

2.22. *Short Walls*

The maximum service moments were calculated previously (Section 2.8) and reference to the Handbook, Table A1.2.1, gives steel requirements as follows:

Horizontal reinforcement on the inner face near the corners ($M = 5{\cdot}72$ kN m/m)
Y12 at 300 c/c gives a moment of resistance of 11·8 kN m/m at a steel stress of 239 N/mm^2. The actual service steel stress is therefore

$$\frac{5{\cdot}72}{11{\cdot}8} \times 239 = 116\,\text{N/mm}^2$$

There now arises the problem of catering for the direct tension. One approach would be to provide additional steel to carry the tensile force at a stress of 116 N/mm^2, i.e. the same stress as is present in the steel provided for bending as suggested in the Handbook, Section 4.3. However, this appears unduly restrictive when the actual service bending stress is less than half the value of 239 N/mm^2 which the table indicates to be acceptable for bending.

In the authors' opinion 239 N/mm^2 is too high having regard to the recommendation in the Handbook, Section A1.4(4). An arbitrary reduction to a maximum of, say, 200 N/mm^2 leaves

$$200 - 116 = 84\,\text{N/mm}^2$$

available to resist direct tension.

On this basis, Y12 at 300 c/c (377 mm^2/m) is capable of resisting direct tension of

$$377 \times \frac{84}{1000} = 31{\cdot}7\,\text{kN/m}$$

which exceeds the calculated total direct tension of 21 kN/m (Section 2.12) and no additional reinforcement is required.

Inner face horiz. Y12 @ 300 c/c

Horizontal reinforcement on the outer face ($M = 9{\cdot}53$ kN m/m)
Y12 at 300 c/c (377 mm^2/m) suffices at a stress of

$$\frac{9{\cdot}53}{11{\cdot}8} \times 239 = 193\,\text{N/mm}^2$$

which is satisfactory.

Additional steel required to carry the direct tension

$$= \frac{21\,000}{193} = 109\,\text{mm}^2/\text{m}$$

Total steel required

$$= 377 + 109 = 486\,\text{mm}^2/\text{m}$$

Outer face Y12 @ 230 c/c

Vertical reinforcement on the inner face ($M = 9{\cdot}60\,\text{kN m/m}$)
Y12 at 300 c/c from the table (at 239 N/mm^2)
Actual stress is

$$239 \times \frac{9{\cdot}60}{11{\cdot}8} \times \frac{250}{300} = 162\,\text{N/mm}^2$$

which is satisfactory.

Inner face vert. Y12 @ 300 c/c (revised to Y12 @ 240 c/c—see Section 2.24)

2.23. *Floor—Short Span*

Maximum mid-span service moment is 24·5 kN m/m as before (Section 2.9) and maximum direct tension is 12 kN/m (Section 2.12). The Handbook, Table A1.2.1, gives Y20 at 250 c/c (1260 mm^2/m) at a steel stress of 165 N/mm^2. At the same stress the additional steel required is

$$\frac{12 \times 1000}{165} = 73\,\text{mm}^2/\text{m}$$

so that the total steel area = 1333 mm^2/m.

Y20 @ 200 c/c

Y20 at 230 c/c would suffice but will be closed to Y20 at 200 c/c to conform with the steel in the long wall for steel fixing purposes.

2.24. *Floor—Long Span*

Maximum mid-span service moment is 6·6 kN m/m (Section 2.11) and the direct tension is taken as 12 kN/m (Section 2.12). For bending, Table A1.2.1 gives Y12 at 300 c/c (377 mm^2/m) for a moment of 11·8 kN m at a steel stress of 239 N/mm^2. Stress at service moment

$$= \frac{6{\cdot}6}{11{\cdot}8} \times 239 = 134\,\text{N/mm}^2$$

Additional steel for the direct tension

$$= \frac{12 \times 1000}{133} = 90\,\text{mm}^2/\text{m}$$

giving a total of

$$377 + 90 = 467\,\text{mm}^2/\text{m}$$

*Y*12 @ 240 *c/c*

This spacing differs from the 300 c/c previously selected for vertical steel in the short wall which will now be revised to 240 c/c for ease of steel fixing.

DESIGN NO. 3: ALTERNATIVE METHOD OF DESIGN—BS 5337, CLAUSES 6, 9 AND 12

2.25. *General Approach*

This method is little altered from the provisions of the earlier code CP 2007: Part 2: 1970. It is based entirely on working loads and each member is designed to comply with the two requirements of Clause 6.2.1, namely

(a) resistance to cracking (Clauses 9.2.1 and 9.3.1) and
(b) strength (Clauses 9.2.2 and 9.3.2)

The values of maximum moments and direct tensions are found as before and are listed below.

(a) Long walls:
  (i) maximum vertical bending moment = 28·6 kN m/m;
  (ii) maximum horizontal tension taken as 26 kN/m.†
(b) Short walls:
  (i) maximum vertical bending moment = 9·6 kN m/m;
  (ii) maximum horizontal bending moment = 5·72 kN m/m (corners) and = −9·53 kN m/m (centres);
  (iii) maximum direct horizontal tension taken as 21 kN/m.†
(c) Floor:
  (i) maximum bending moment = 6·6 kN m/m (long span) and = 24·5 kN m/m (short span);
  (ii) maximum direct tension = 12 kN/m.

† In fact the section properties used previously will change slightly due to the changed wall and floor thicknesses. Nevertheless the difference is small and retention of these values simplifies comparisons.

Clearly the maximum moment occurs in the vertical direction in the long walls and this will be considered first.

2.26. *Long Walls*

Try a section 200 mm thick, $d = 140$ mm, concrete Grade 30 and deformed high yield reinforcement as before. Consider the resistance to cracking. Section modulus of uncracked concrete Z_c

$$= \frac{1000 \times 200^2}{6} = 6{\cdot}67 \times 10^6 \text{ mm}^3$$

Tensile stress due to bending

$$= \frac{28{\cdot}6 \times 10^6}{6{\cdot}67 \times 10^6} = 4{\cdot}3 \text{ N/mm}^2$$

which is more than double the permissible value of $2{\cdot}02 \text{ N/mm}^2$ given in the Code, Table 2.

It is, however, possible to include the effect of the steel reinforcement in calculating the effective concrete tensile stress. As a first approximation assume 1% steel in each face. The modular ratio is taken as 15 in accordance with the Code, Clause 9.1(b).

The effective area of concrete then becomes as in Fig. 2.5. Transforming the steel to an equivalent area of concrete involves multiplying its area by the modular ratio of 15 but note that since the 'holes' previously occupied by the bars are now imagined to be concrete, the added area of concrete is $15 - 1 = 14$ times the steel area. The moment of inertia of the transformed steel is given by

$$\frac{bh^3}{12} + \left[14 \times A_s \times \left(d - \frac{h}{2}\right)^2\right]$$

$$= \frac{1000 \times 200^3}{12} + [14 \times (2 \times 0{\cdot}01 \times 200.000)(140 - 100)^2]$$

$$= 756 \times 10^6 \text{ mm}^4$$

and

$$Z_c = \frac{756 \times 10^6}{100} = 7{\cdot}56 \times 10^6 \text{ mm}^3$$

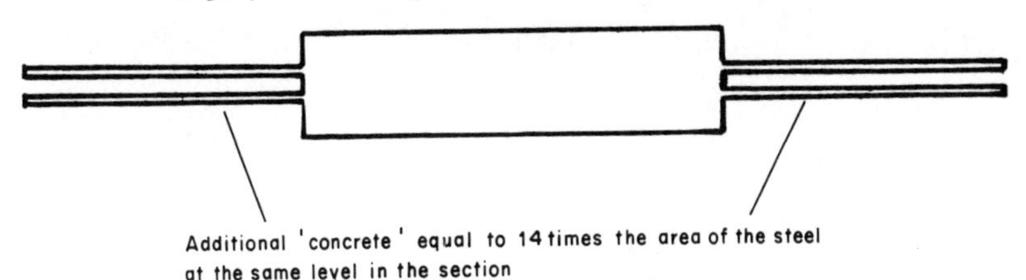

Fig. 2.5. *Transformed section.*

Maximum concrete stress

$$= \frac{28{\cdot}6 \times 10^6}{7{\cdot}56 \times 10^6} = 3{\cdot}78\,\text{N/mm}^2$$

which is still too high.

It may be seen that the more accurate assessment of concrete tensile stress, including an allowance for the small area of steel required for strength purposes, gives a reduction in stress only of the order of 10 % when compared with the approximate calculation on the concrete alone.

In this instance the section must now be increased. Try $h = 280\,\text{mm}$, $d = 220\,\text{mm}$. For the concrete alone

$$Z_c = \frac{1000 \times 280^2}{6} = 13{\cdot}07 \times 10^6\,\text{mm}^3$$

Tensile stress due to bending

$$= \frac{28{\cdot}6 \times 10^6}{13{\cdot}07 \times 10^6} = 2{\cdot}2\,\text{N/mm}^2$$

This is marginally too high and will be re-checked when the reinforcement is known.

Consider the strength requirement:

$$\frac{M}{bd^2} = \frac{28{\cdot}6 \times 10^6}{1000 \times 220^2} = 0{\cdot}59$$

The Code, Clause 9.1, requires elastic design and the design chart (Fig. 1.1) gives

$$p = 0{\cdot}005 \text{ (with } f_c = 4{\cdot}2\,\text{N/mm}^2\text{)}$$
$$A_s = 0{\cdot}005 \times 220 \times 1000 = 1100\,\text{mm}^2/\text{m}$$

See Section 2.29

Table 2.2

Item	*Area* (A), mm^2	'y' *to tension face, mm*	Ay $\times 10^6$	I $\times$ 10^9 mm^4	A$\bar{y}$ $\times 10^9$
Concrete	280 000	140	39·2	1·83	—
Transformed steel	15 400	60	0·9	—	0·09
	295 400		40·1		

The transformed section properties may be calculated as in Table 2.2.

$$\bar{y} = \frac{40{\cdot}1 \times 10^6}{294\,000} = 136\,\text{mm}$$

$$I = 1{\cdot}92 \times 10^9\,\text{mm}^4$$

The section modulus for the tension face Z

$$= \frac{1{\cdot}92 \times 10^9}{136} = 14{\cdot}12 \times 10^6\,\text{mm}^3$$

and tensile stress in concrete

$$= \frac{28{\cdot}6 \times 10^6}{14{\cdot}12 \times 10^6} = 2{\cdot}02\,\text{N/mm}^2$$

which is just permissible.

In the horizontal direction, steel of area 0·15 % will be used in each face. Check direct tension effect: Effective area of concrete

$$= 280\,000 + \left(2 \times \frac{0{\cdot}15}{100} \times 14 \times 280\,000\right)$$

$$= 291\,760\,\text{mm}^2$$

Tensile stress on concrete

$$= \frac{21 \times 10^3}{291\,760} = 0{\cdot}07\,\text{N/mm}^2$$

which is satisfactory (permissible 1·44 N/mm² from Code, Table 2). Above the kicker the required moment is 9·9 kN/m as before (Section 2.7). The tensile stress is obviously satisfactory and the strength calculation only is required.

$$\frac{M}{bd^2} = \frac{9{\cdot}9 \times 10^6}{1000 \times 220^2} = 0{\cdot}20$$

$$p = 0{\cdot}0015$$
$$A_s = 330\,\text{mm}^2/\text{m}$$

See Section 2.29

2.27. *Short Walls*

Try $h = 280$, $d = 220$ for vertical steel and $d = 235$ for horizontal steel.

Vertical steel

$$\frac{M}{bd^2} = \frac{9{\cdot}6 \times 10^6}{1000 \times 220^2} = 0{\cdot}20$$

$$p = 0{\cdot}002$$
$$A_s = 440\,\text{mm}^2/\text{m}$$

See Section 2.29

Horizontal steel (*corners*)

$$\frac{M}{bd^2} = \frac{5{\cdot}72 \times 10^6}{1000 \times 235^2} = 0{\cdot}10$$

Take $p = 0{\cdot}0015$ (minimum to Clause 4.11.2 of the Code):

$$A_s = 353\,\text{mm}^2$$

See Section 2.29

Horizontal steel (*centre*)

$$\frac{M}{bd^2} = \frac{9{\cdot}53 \times 10^6}{1000 \times 235^2} = 0{\cdot}17$$

$$p = 0{\cdot}002$$
$$A_s = 470\,\text{mm}^2/\text{m}$$

See Section 2.29

Check maximum tensile stress:

$$Z_c = \frac{1000 \times 280^2}{6} = 13{\cdot}07 \times 10^6\,\text{mm}^3$$

Tensile stress due to bending

$$= \frac{9{\cdot}53 \times 10^6}{13{\cdot}07 \times 10^6} = 0{\cdot}73\,\text{N/mm}^2$$

Tensile stress due to direct tension

$$= \frac{21 \times 10^3}{280 \times 1000} = 0{\cdot}08\,\text{N/mm}^2$$

Total tension stress $f_{ct} = 0{\cdot}81\,\text{N/mm}^2$ which is satisfactory.

2.28. *Floor*

Take $h = 280$ mm and $d = 220$ mm (i.e. as for walls). The increased self-weight increases the previous bending moments by 3 kN m (short span) and 1·1 kN m (long span).

Short Span

$$\frac{M}{bd^2} = \frac{27{\cdot}5 \times 10^6}{1000 \times 220^2} = 0{\cdot}57$$

$$p = 0{\cdot}0049$$
$$A_s = 1078\,\text{mm}^2$$

See Section 2.29

Check maximum tensile stress: Tensile stress due to bending

$$= \frac{27{\cdot}5 \times 10^6}{13{\cdot}07 \times 10^6} = 2{\cdot}10\,\text{N/mm}^2$$

Tensile stress due to direct tension

$$= \frac{12 \times 10^3}{280 \times 1000} = 0{\cdot}04\,\text{N/mm}^2$$

Total $f_{ct} = 2{\cdot}14\,\text{N/mm}^2$ which exceeds the permissible value of $2{\cdot}02\,\text{N/mm}^2$. However, a calculation which allows for the presence of the reinforcement (as for the long walls, Section 2.26) yields a value of f_{ct} less than $2{\cdot}02\,\text{N/mm}^2$.

Long Span

$$\frac{M}{bd^2} = \frac{7{\cdot}7 \times 10^6}{1000 \times 235^2} = 0{\cdot}14$$

Take minimum $p = 0{\cdot}0015$:

$$A_s = 353\,\mathrm{mm}^2$$

See Section 2.29

Tensile stress is obviously satisfactory by comparison with short span.

2.29. *Reinforcement Arrangement*

Summarising the calculated requirements for the alternative method of design:

Long walls—vertical steel at base = $1100\,\mathrm{mm}^2/\mathrm{m}$

*Y*20 @ 280 *c*/*c*

Long walls—vertical steel at kicker = $330\,\mathrm{mm}^2/\mathrm{m}$

*Y*12 @ 280 *c*/*c*

Floor—short span (bottom) = $1078\,\mathrm{mm}^2/\mathrm{m}$

*Y*20 @ 280 *c*/*c*

Short walls—vertical steel = $440\,\mathrm{mm}^2/\mathrm{m}$

*Y*12 @ 250 *c*/*c*

Floor—long span (bottom) = $353\,\mathrm{mm}^2/\mathrm{m}$

*Y*12 @ 250 *c*/*c*

Short walls—corner horizontal (inner) = $353\,\mathrm{mm}^2/\mathrm{m}$

*Y*12 @ 240 *c*/*c*

Short walls—centre horizontal (outer) = $470\,\mathrm{mm}^2/\mathrm{m}$

*Y*12 @ 240 *c*/*c*

(All other faces $353\,\mathrm{mm}^2/\mathrm{m}$)

*Y*12 @ 250 *c*/*c*

Listing of these requirements in the pairs as above facilitates consideration of steel fixing requirements in relation to the 'opening corner' detail described in Chapter 1. The bar diameters and spacings can now be

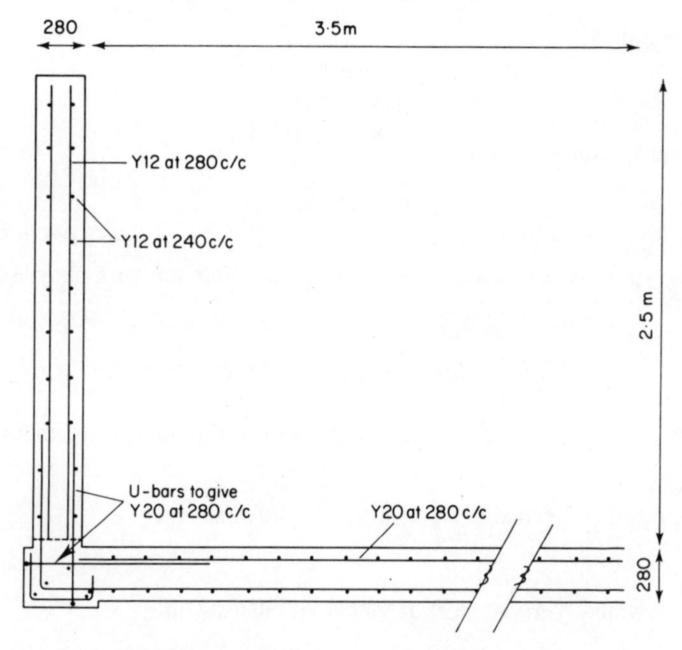

Fig. 2.6. *Reinforcement for Design No. 3.*

entered in the right-hand output margin as above. Note that in practice 20-mm bars are the minimum size for the main steel in a floor slab. Inevitably this leads to some over-provision of reinforcement but smaller sizes will almost certainly be displaced to an unacceptable extent during concrete placing. The authors consider maximum bar spacing should not exceed 300 mm in these structures although the Code does not modify CP 114 in this respect. Clause 308 of CP 114 would permit main bars in this tank to be spaced at up to 660 mm and distribution bars at up to 1100 mm. These spacings must be regarded as excessive in conditions where crack control is of paramount importance.

The increase in wall and floor thickness from 200 mm for the limit state designs to 280 mm for the alternative method would have permitted design of the outer face to Class C exposure (Code, Clause 4.9). Hence the design for sagging moments in the floor could have been on the basis of CP 114 without consideration of cracking. The calculations have, however, been prepared for Class B exposure to illustrate the method. The reinforcement details can now be drawn as in Fig. 2.6, and this may be compared with Fig. 2.4.

CHAPTER 3

DESIGN OF AN 'IN GROUND' REINFORCED CONCRETE SWIMMING POOL

DESIGN INFORMATION

3.1. *Design Codes*

BS 5337: 1976 (referred to as 'the Code').
CP 110: 1972.

3.2. *Other Design Data*

Handbook to BS 5337: 1976 (referred to as 'the Handbook').

3.3. *Loading Conditions*

Water—density equivalent to 9·8 kN/m^3.
Reinforced concrete—density equivalent to 25 kN/m^3.
Soil backfill—density equivalent to 20 kN/m^3.
Soil pressure (active), $k_a = \tan^2 (45 - \phi/2)$.

3.4. *Client's Requirements*

The internal dimensions of the pool are as shown in Fig. 3.1. Note: the pool is not required for diving.

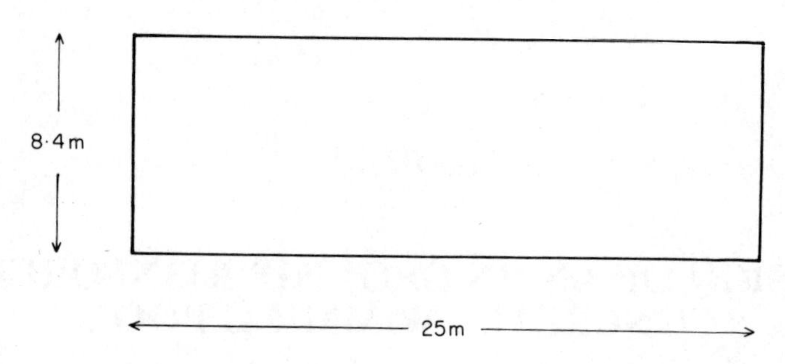

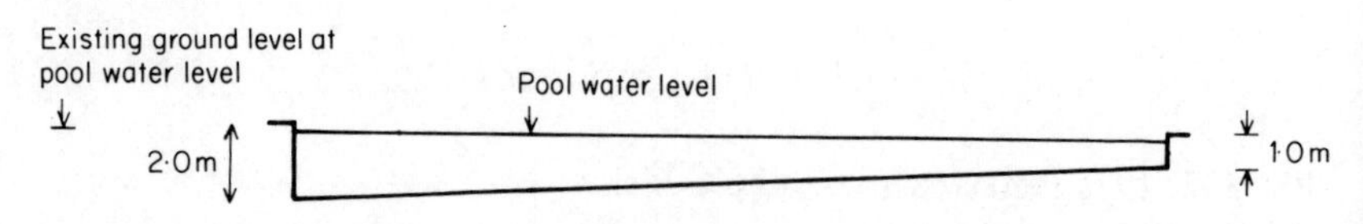

Fig. 3.1. *Internal dimensions of pool.*

3.5. *Topography*

Level site.

3.6. *Borehole Data*

Table 3.1

Description	*Depth below ground level*	N (*Blows Standard Penetration Test*)
Top soil	0–0·4	
Sand with some gravel	0·4–1·8	45
Boulder clay	1·8–8·0+	Cohesion 50 kN/m^2

Water table 2·3 m below ground level (highest recorded level winter or summer).

Density of sand and gravel (moist) = 20 kN/m^3.

3.7. *Exposure Conditions*

Reference Clause 4.9 of BS 5337:
internal surfaces of pool—Class B exposure,
external surfaces of pool—Class C exposure.
Consider top internal 200 mm of wall also as Class B for the reasons given in Chapter 2.
Note: if the walls or slab when designed are 225 mm thick or less, both faces then to be considered Class B.

3.8. *Design Method*

Design to be in accordance with Clause 5 of the Code, i.e. limit state.

3.9. *Design Principles*

Design of the wall and heel to be prepared for two cases: (1) water pressure acting on pool structure with no allowance for reaction from backfill; and (2) pressure of backfill acting on pool structure when empty.
Concrete—Grade 25.
Reinforcement—hot rolled or cold worked high yield deformed.

DESIGN CALCULATIONS

3.10. *Deflection*

Try a wall thickness of 225 mm: check for deflection under Clause 5.3.1. Basic span/effective depth ratio for triangular load distribution on cantilever walls (applying 25% increase—see Handbook)

$$= 1{\cdot}25 \times 7 = 8{\cdot}75$$

Assuming a modification factor of 1·5 for tension reinforcement, the maximum span/effective depth

$$= 1{\cdot}5 \times 8{\cdot}75 = 13{\cdot}1$$

Cover 40 mm, distribution steel, say, 8-mm diameter. Hence for 10-mm main bars

$$d = 225 - 40 - 8 - 5 = 172\,\text{mm}$$

The actual span/effective depth ratio

$$= \frac{2000}{172} = 11{\cdot}6$$

Satisfactory

The wall will be reinforced on each face, so for any one load condition both tensile and compressive reinforcement will be present. A further modification factor for compressive reinforcement could therefore be used. However, there will be a tendency for the heel to rotate slightly under load such that the deflection at the top of the wall will increase. For this reason it is always better to keep the actual span/effective depth ratio well below the maximum allowable in cantilever wall construction.

For this pool size it will not be economical to reduce the wall thickness as the water depth reduces or to taper the back face of the wall.

3.11. *Stability* (*Full*)

Check the stability per metre length of wall at maximum depth only (assume the pool is filled to 1·8 m depth):

Referring to Fig. 3.2

Weight of water on heel $W_1 = 2{\cdot}0 \times 1{\cdot}8 \times 9{\cdot}8 = 35{\cdot}3\,\text{kN}$

Weight of wall $W_2 = 0{\cdot}225 \times 2{\cdot}225 \times 25 = 12{\cdot}5\,\text{kN}$

Weight of heel $W_3 = 2{\cdot}0 \times 0{\cdot}225 \times 25 = 11{\cdot}2\,\text{kN}$

Force due to water pressure $P = \dfrac{9{\cdot}8 \times 1{\cdot}8^2}{2} = 15{\cdot}9\,\text{kN}$

Taking moments about point A:

overturning moment $= 15{\cdot}9 \times 0{\cdot}825 = 13{\cdot}1\,\text{kN m}$

restoring moment $= (35{\cdot}3 \times 1{\cdot}225) + (12{\cdot}5 \times 0{\cdot}112)$
$+ (11{\cdot}2 \times 1{\cdot}225) = 58{\cdot}4\,\text{kN m}$

Therefore the factor of safety against overturning

$$= \frac{58{\cdot}4}{13{\cdot}1} = 4{\cdot}4$$

Adequate

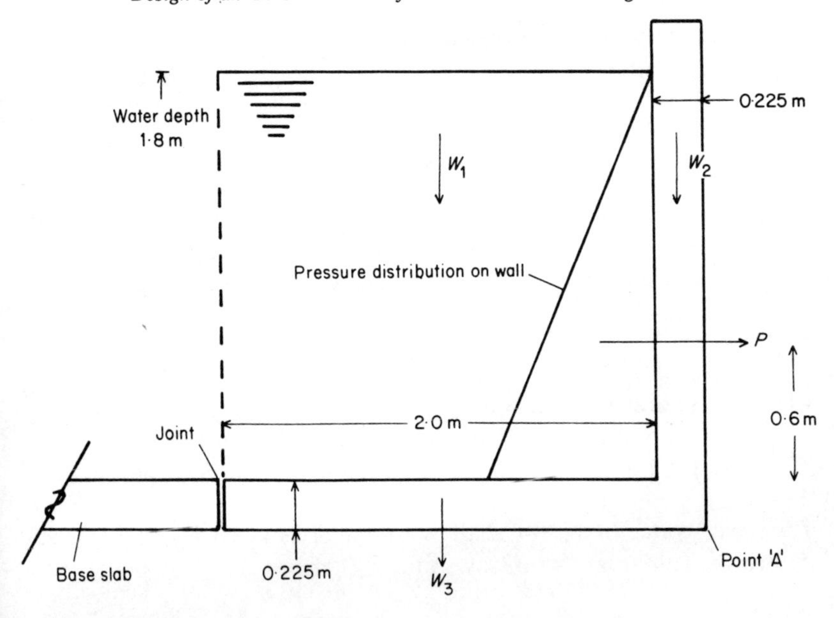

Fig. 3.2. *Forces on the pool wall when full.*

3.12. *Bearing Pressures* (*Full*)

Bearing pressures on the underside of the heel: centroid of vertical forces

$$W_1 + W_2 + W_3 = 59{\cdot}0\,\text{kN}$$

$$59{\cdot}0 \times \bar{x} = 58{\cdot}4 \text{ (}\bar{x}\text{ measured from point A)}$$

$$\bar{x} = 0{\cdot}990\,\text{m}$$

Heel length = 2·225 m
Eccentricity of applied force = 0·122 m
Moment of forces about centroid of base

$$= (59{\cdot}0 \times 0{\cdot}122) + 13{\cdot}1 = 20{\cdot}3\,\text{kN m}$$

Z of base

$$= 1{\cdot}0 \times \frac{2{\cdot}225^2}{6} = 0{\cdot}83\,\text{m}^3$$

Therefore the pressure on the clay

$$= \frac{59{\cdot}0}{2{\cdot}225} \times 1{\cdot}0 \pm \frac{20{\cdot}3}{0{\cdot}83} = 26{\cdot}5 \pm 24{\cdot}4$$

$$= 50{\cdot}9 \text{ to } 2{\cdot}1 \text{ kN/m}^2$$

Check the allowable soil pressures (for more detailed treatment see the text book by M. J. Tomlinson included in the bibliography at the end of this chapter). Considering the clay, the ultimate bearing capacity q

$$= cN_c + \gamma D$$

where c = undrained cohesion,
N_c = bearing capacity factor,
γD = total overburden pressure at foundation level.

The lowest value of N_c for shallow foundations given by Terzaghi and quoted by Tomlinson for a clay soil is 5·7. Hence

$$q = 5{\cdot}7 \times 50 + 20 \times 2 = 325 \text{ kN/m}^2$$

The actual maximum bearing pressure = 50·9 kN/m². Therefore the factor of safety

$$= \frac{325}{50{\cdot}9} = 6{\cdot}3$$

Satisfactory

Since the overburden pressure is only slightly less than the maximum ground pressure under the structure, settlements will be very small. Considering the sand stratum and referring to the familiar Terzaghi and Peck curves (also reproduced in the text book referred to above) it is found that for a value of $N = 45$ (see Section 3.6) and foundation width 2·25 m the allowable bearing pressure is 460 kN/m² but this value should be halved for a water table within 2·25 m of the foundation level. Hence

$$\text{allowable bearing pressure} = 230 \text{ kN/m}^2$$

The maximum calculated stress is 50·9 kN/m² and the factor of safety is clearly adequate by inspection. Had the factor of safety been lower then the fact that the sand stratum will only carry the shallower parts of the pool could have been considered.

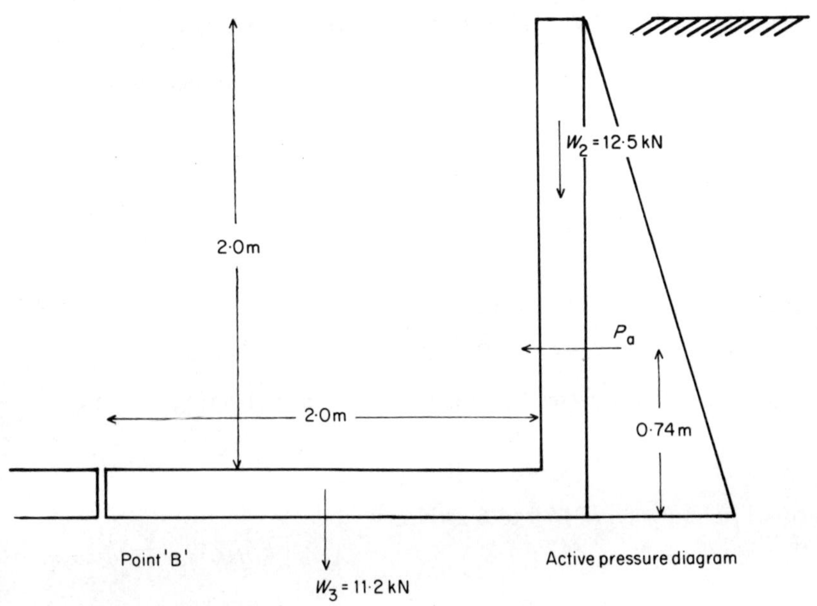

Fig. 3.3. *Forces on the pool wall when empty.*

3.13. *Stability (Empty)*

Pressures on the wall due to retained sand and gravel acting on the empty pool are shown in Fig. 3.3.
Total height of retained material $H = 2{\cdot}225\,\text{m}$
For $N = 45$, ϕ of sand and gravel is approximately 40°.

$$k_a = \tan^2\left(45 - \frac{\phi}{2}\right) = 0{\cdot}217$$

Active pressure P_a

$$= k_a \frac{W_g H^2}{2} = 0{\cdot}217 \times \frac{20 \times 2{\cdot}225^2}{2} = 10{\cdot}7\,\text{kN}$$

acting at $H/3 = 0{\cdot}74\,\text{m}$ above base.

Taking moments about point B:

overturning moment $= 10{\cdot}7 \times 0{\cdot}74 = 7{\cdot}9\,\text{kN m}$

restoring moment $= (11{\cdot}2 \times 1{\cdot}0) + (12{\cdot}5 \times 2{\cdot}112) = 37{\cdot}6\,\text{kN m}$

Therefore the factor of safety against overturning $= 4{\cdot}8$.

Adequate

3.14. *Bearing Pressures (Empty)*

Bearing pressures at underside of heel: centroid of vertical forces

$$W_3 + W_2 = 23{\cdot}7\,\text{kN}$$

Assume centroid $\bar{x}$ from point B:

$$37{\cdot}6 = 23{\cdot}7\bar{x}$$
$$\bar{x} = 1{\cdot}6\,\text{m}$$

and the resultant moment on base about its centroid 1·112 m from point B

$$= (1{\cdot}6 - 1{\cdot}112) \times 23{\cdot}7 - 7{\cdot}9 = 3{\cdot}66\,\text{kN m}$$

Z of base $= 0{\cdot}83\,\text{m}^3$ (see Section 3.12). Therefore bearing pressure

$$= \frac{23{\cdot}7}{2{\cdot}225 \times 1{\cdot}0} \pm \frac{3{\cdot}66}{0{\cdot}83} = 10{\cdot}6 \pm 4{\cdot}4$$
$$= 15{\cdot}0 \text{ to } 6{\cdot}2\,\text{kN/m}^2$$

Satisfactory (no tension)

3.15(*a*). *Reinforcement in Wall/Water Face (Using Code Table* 1)

Outward moment due to water pressure (Section 3.11)

$$= 15{\cdot}9 \times 0{\cdot}6 = 9{\cdot}54\,\text{kN m}$$

Steel stress not to exceed values given in Table 1 of the Code, i.e. allowable stress

$$= 130\,\text{N/mm}^2 \text{ (for water face)}$$

$$\frac{M}{bd^2} = \frac{9{\cdot}54 \times 10^6}{10^3 \times 172^2} \qquad (d = 172\,\text{mm as in Section 3.10})$$
$$= 0{\cdot}32$$

$$\frac{A_s}{bd} = 0{\cdot}0027$$

The modification factor of 1·5 assumed in Section 3.10 is therefore justified—actual value 2·0.

$$A_s = 0{\cdot}0027 \times 10^3 \times 172 = 467\,\text{mm}^2/\text{m}$$

The Code does not give guidance on the maximum reinforcement spacing to be used in conjunction with the stresses in Table 1. However, it is prudent to use small diameter bars at centres not exceeding about 300 mm. Hence try Y12 at 240 c/c (471 mm^2/m).

Y12 @ 240 c/c

Check Ultimate Capacity of Section
Ultimate applied moment

$$= 1{\cdot}6 \times 9{\cdot}54 = 15{\cdot}3\,\text{kN m}$$

Referring to Chart 2 of CP 110

$$\frac{100A_s}{bd} = \frac{100 \times 471}{10^3 \times 172} = 0{\cdot}27$$

$$\frac{M}{bd^2} = 0{\cdot}9$$

Therefore

$$M = 0{\cdot}9 \times \frac{10^3}{10^6} \times 172^2 = 26{\cdot}9\,\text{kN m}$$

Adequate

Check Ultimate Shear Stress

$$v = \frac{15{\cdot}9 \times 1{\cdot}6 \times 10^3}{10^3 \times 172} = 0{\cdot}15\,\text{N/mm}^2$$

Allowable shear stress = 0·35 N/mm^2 (Table 5, CP 110).

Satisfactory

3.15(*b*). *Reinforcement in Wall/Water Face (Alternative with Crack Width Calculated)*

The authors do not recommend that the steel stresses are permitted to approach the high values in the order of 240 N/mm^2 which can be shown to comply with the Code when using this method. The limiting value to be used is a matter of judgement for the designer and it is possible that the higher values will be justified by experience. For this example the authors suggest a limit of $180\,\mathrm{N/mm^2}$ and this value will be applied here.

$$\frac{M}{bd^2} = \frac{9{\cdot}54 \times 10^6}{1000 \times 172^2} = 0{\cdot}32$$

From Fig. 1.1

$$\frac{A_s}{bd} = 0{\cdot}0019$$

$$A_s = 0{\cdot}0019 \times 1000 \times 172 = 327\,\mathrm{mm^2/m}$$

Y10 @ 240 c/c

For this particular design the required area of reinforcement exactly equals that provided. In the more general case the practical spacing of bars is such that a small excess is provided. The value of f_{st} used in the subsequent calculation could then have been reduced to

$$f_{st} = \frac{\text{(required area)}}{\text{(area provided)}} \times 180\,\mathrm{N/mm^2}$$

From Fig. 1.1

$$\frac{x}{d} = 0{\cdot}21$$

Therefore

$$x = 0{\cdot}21 \times 172 = 36$$

The dimensions can now be entered on Fig. 3.4 as shown.

Taking $E_s = 200\,\mathrm{kN/mm^2}$ (CP 110, Clause 2.4.2.4):

$$\varepsilon_s = \frac{180}{200 \times 10^3} = 0{\cdot}000\,9$$

Therefore

$$\varepsilon_1 = \frac{189}{136} \times 0{\cdot}0009 = 0{\cdot}001\,25$$

The Code, Appendix C, equation (2) gives

$$\varepsilon_m = 0{\cdot}001\,25 - \frac{0{\cdot}7 \times 1000 \times 225(225 - 36)}{327(225 - 36)180 \times 10^3} = -0{\cdot}0014$$

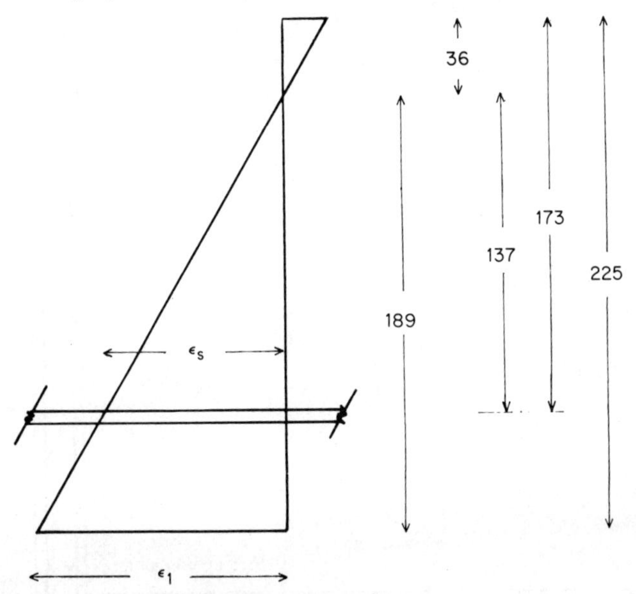

Fig. 3.4. *Strain diagram.*

Since the negative sign indicates an uncracked section (Code, Appendix C) no further calculation is required.

Satisfactory

Check M_u:

$$\frac{100A_s}{bd} = 0{\cdot}19$$

From Chart 2 of CP 110

$$\frac{M}{bd^2} = 0{\cdot}7$$

Therefore

$$M_u = \frac{0{\cdot}7 \times 10^3 \times 172^2}{10^6}$$
$$= 20{\cdot}7\,\text{kN m.}$$

(Required $M_u = 15{\cdot}3\,\text{kN m.}$)

Adequate

The reinforcement details can now be drawn as shown in Fig. 3.5.

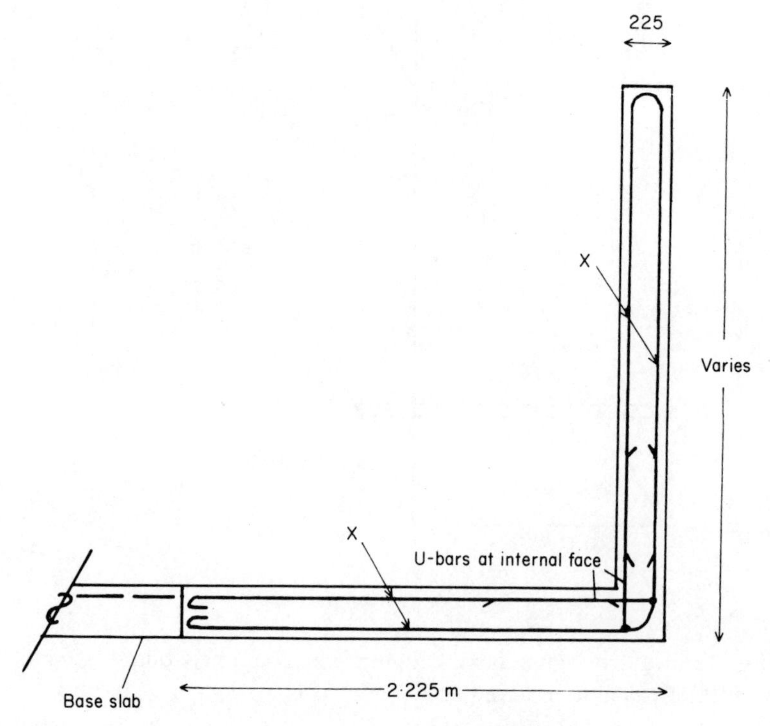

Fig. 3.5. *Alternatives for main reinforcement: (a) using allowable stresses to Code Table 1 for which bars X are Y12 @ 240 c/c and (b) using calculated crack widths for which bars X are Y10 @ 240 c/c.*

3.15(*c*). *Reinforcement in Wall/Water Face (Alternative Using Tables in Handbook, Appendix* 1)

As before

$$M_d = 9{\cdot}54\,\text{kN m}$$

Table A1.2.1 gives Y12 at 300 c/c for $M = 11{\cdot}8$ and $f_{st} = 239$ and $h = 200$.

Note that although there is an excess moment capacity the bar spacing cannot necessarily be increased without invalidating the crack width limit. However, the actual f_{st} may be estimated by interpolation between Tables A1.2.1 ($h = 200$) and A1.2.2 ($h = 250$), i.e.

$$h = 200$$

$$f_{st} = \frac{9{\cdot}54}{11{\cdot}8} \times 239 = 193\,\text{N/mm}^2$$

and

$$h = 250$$

$$f_{st} = \frac{9{\cdot}54}{15{\cdot}9} \times 236 = 142\,\text{N/mm}^2$$

Hence for $h = 225$

$$f_{st} \simeq 168\,\text{N/mm}^2$$

Satisfactory

3.16. *Reinforcement in Wall/Earth Face*

Design of earth face (consider as Class B since the wall is only 225 mm thick).
M due to earth pressure at base of stem

$$= 0{\cdot}217 \times 20 \times \frac{2^3}{6} = 5{\cdot}79\,\text{kN m}$$

$$\frac{M}{bd^2} = \frac{5{\cdot}79 \times 10^6}{10^3 \times 172^2} = 0{\cdot}196$$

The Code, Table 1, gives $f_{st} = 130\,\text{N/mm}^2$ for method (a) as in Section 3.15(a).

From Fig. 1.1

$$100\frac{A_s}{bd} = 0{\cdot}15$$

$$A_s = 0{\cdot}15 \times \frac{10^3}{100} \times 172 = 258\,\text{mm}^2/\text{m}$$

Minimum percentage of reinforcement permitted, high yield

$$= \frac{0{\cdot}15}{100} \times 10^3 \times 225 = 337\,\text{mm}^2/\text{m}$$

Therefore provide minimum of 337 mm^2/m.

Y10 @ 225 c/c

Method (b) cannot therefore be considered and method (c) will result in a higher steel content as in Section 3.15 (c).

3.17. *Heel*

The above steel areas in each face of the wall will also be provided in the appropriate face of the heel (since the maximum moment equals that in the wall).

3.18. *Wall Corners*

Referring to Clause 4.3 of the Code it will be noted that the walls have been designed as free-standing cantilevers; the small amounts of direct tension induced in a horizontal plane in the walls at their corners will be dealt with by the provision of extra detailing steel.

3.19. *Temperature and Moisture Effects*

There is no restraint to wall contraction in a vertical direction. Similarly there is no restraint to the heel in a direction perpendicular to the wall. Continuous construction is planned for the walls which will be constrained in the longitudinal direction against temperature and moisture shrinkage effects. The designer is given two choices in the Code under Clauses 4.11.1 and 4.11.2. Nominal reinforcement under Clause 4.11.2 may be provided and no check on crack widths is then required. However, if this option is used no guarantee can be given that the crack widths will be satisfactory. It is proposed in this design to check the crack widths using Appendix B and Clauses 4.11.1 and 15.1 of the Code.

Wall in horizontal plane: Appendix B of the Code, Clause B1, gives minimum reinforcement

$$\rho_{crit} = \frac{f_{ct}}{f_y}$$

For Grade 25 concrete

$$f_{ct} = 1{\cdot}15\,\mathrm{N/mm^2}$$

and

$$f_y = 425\,\mathrm{N/mm^2}$$

Therefore

$$\rho_{crit} = \frac{1{\cdot}15}{425} = 0{\cdot}0027$$

Therefore area of reinforcement required

$$= 0{\cdot}0027 \times 225 \times 1000 = 608\,\mathrm{mm^2}$$

i.e. $304\,\mathrm{mm^2}$ each face.

Try Y10 at 250 mm c/c ($314\,\mathrm{mm^2}$) each face. Then

$$\rho = \frac{314 \times 2}{225 \times 10^3} = 0{\cdot}002\,79$$

Check crack width using Clause B2†

$$S_{max} = \frac{f_{ct}}{f_b} \times \frac{\phi}{2\rho_c}$$

$$= \frac{4}{5} \times \frac{10}{2 \times 0{\cdot}002\,79} \quad \text{(using Type 1 bars)}$$

$$= 1433\,\text{mm}$$

$$W_{max} = S_{max} \times \frac{\alpha}{2} \times (T_1 + T_2)$$

T_1 = for concreting in summer = 30 °C
T_2 = 20 °C (say)
$\alpha = 12 \times 10^{-6}$ per °C

$$W_{max} = 1433 \times \frac{12 \times 10^{-6}}{2} \times (50) = 0{\cdot}43\,\text{mm}$$

which is excessive (maximum for Class B is 0·2 mm).

Find S_{max} required:

$$S_{max} = \frac{0{\cdot}2}{0{\cdot}43} \times 1433 = 666\,\text{mm}$$

Substituting into

$$S_{max} = \frac{f_{ct}}{f_b} \times \frac{\phi}{2\rho_c}$$

required $\phi/2\rho_c$

$$= 666 \times \frac{5}{4} = 833$$

Using 10-mm bars

$$\rho_c = \frac{10}{2 \times 833} = 0{\cdot}006$$

i.e. 0·003 each face.

$$A_s \text{ (each face)} = 0{\cdot}003 \times 225 \times 10^3 = 675\,\text{mm}^2/\text{m}$$

Y10 @ 110 c/c

† Note that whilst the Handbook uses ρ_c, the Code uses p for the steel ratio based on gross concrete section.

or using 8-mm bars

$$\rho_c = \frac{8}{2 \times 833} = 0{\cdot}0048$$

$$A_s \text{ (each face)} = 0{\cdot}0024 \times 225 \times 10^3 = 540\,\text{mm}^2/\text{m}$$

Alternative, Y8 @ 90 c/c

In practice the 10-mm bars would probably be preferred to reduce fixing costs and for greater rigidity during concrete placing. Nevertheless a 20% reduction in distribution steel could represent a worthwhile economy in favour of the 8-mm bars.

3.20. *Temperature and Moisture Effects—Alternative Approach*

The direct use of the Code formulae from Appendix B results in a lengthy calculation as seen in Section 3.19. An alternative is to combine the two requirements for S_{max} as

$$S_{max} = \frac{f_{ct}}{f_b} \times \frac{\phi}{2\rho_c}$$

and

$$S_{max} = \frac{W_{max}}{\frac{1}{2}\alpha(T_1 + T_2)}$$

Equating gives

$$\rho_c = \frac{f_{ct}}{f_b} \times \frac{\phi(T_1 + T_2) \times \alpha}{4W_{max}}$$

Substituting the values used in Section 3.19 for 10-mm bars gives

$$\frac{4}{5} \times \frac{10(50) \times 12 \times 10^{-6}}{4 \times 0{\cdot}2} = 0{\cdot}006, \text{ etc.}$$

as before.

ALTERNATIVE DESIGN WITH MOVEMENT JOINTS

3.21. *Code Option* 2 *and Proposed New Option* 3

The above method for calculating the horizontal steel is in accordance with Design Option 1 in Table 6 of the Code. Other options are available, the most relevant ones being Option 2 from the Code and the proposed new Option 3 which appears in the Handbook in Table 15.1(d) and which, it is suggested, should replace Options 3, 4 and 5 in the Code.

Option 2
Table 6 of the Code allows semi-continuous construction as a method of control under this option. Movement joint spacings of 7·5 m for partial contraction joints or 15·0 m for complete contraction joints are required. In this case a complete contraction joint at the centre of the pool will be used. The minimum steel ratio required under this option is ρ_{crit}. From Section 3.20:

$$\rho_c = \frac{f_{ct}}{f_b} \times \frac{\phi \times T_1 \times \alpha}{4W_{max}} \tag{1}$$

(Note that Appendix B2 of the Code allows the effect of T_2 to be ignored when full movement joints are provided at not more than 15 m centres.) And from Clause B1 of the Code

$$\rho_{crit} \not< \frac{f_{ct}}{f_y} \tag{2}$$

Equation (1) gives the following using values as before except that only T_1 is now considered:

$$\rho_c = \frac{4}{5} \times \frac{12 \times 10^{-6} \times \phi \times 30}{4 \times 0{\cdot}2}$$
$$= 0{\cdot}000\,36\phi$$

Therefore for 10-mm bars

$$\rho_c = 0{\cdot}0036$$

and for 8-mm bars

$$\rho_c = 0{\cdot}0029$$

and for 12-mm bars

$$\rho_c = 0{\cdot}0043$$

Equation (2) gives

$$\rho_{crit} \nless \frac{1{\cdot}15}{425} \qquad \text{(Grade 25 concrete)}$$
$$= 0{\cdot}0027$$

which is less than the values above.

Therefore, using 10-mm-diameter bars

$$A_s \text{ required} = 0{\cdot}0036 \times 225 \times 1000 = 810\,\text{mm}^2/\text{m}$$

(i.e. 405 mm^2/m each face)

Y10 @ 190 c/c both faces

or using 8-mm-diameter bars

$$A_s \text{ required} = 0{\cdot}0029 \times 225 \times 1000 = 653\,\text{mm}^2/\text{m}$$

or using 12-mm-diameter bars

$$A_s \text{ required} = 0{\cdot}0043 \times 225 \times 1000 = 968\,\text{mm}^2/\text{m}$$

Proposed New Option 3

This is given in Table 15.1(d) of the Handbook. Close joint spacing is required and the spacing of the joints must conform with one of the three choices (a), (b) or (c) given in the table. Again the effect of T_2 may be neglected in accordance with Appendix B2 of the Code.

The minimum ratio of steel required is $\frac{2}{3}\rho_{crit}$.

$$\tfrac{2}{3}\rho_{crit} = \tfrac{2}{3} \times 0{\cdot}0027 \qquad \text{(from calculations as in Option 2)}$$
$$= 0{\cdot}0018$$

The Handbook does not include reference to the minimum ratio of 0·002 (0·2%) which appears in the Code for Options 4 and 5. Hence

$$A_s \text{ required} = 0{\cdot}0018 \times 225 \times 1000 = 405\,\text{mm}^2/\text{m}$$

The Handbook indicates that the smaller diameter bars should be avoided for Cases (b) and (c) below. The effect is to increase the value of S_{max} and hence to widen the joint spacing. This suggestion of increased joint spacing for larger diameter bars is contrary to the experience of many practising engineers. In the authors' opinion it should be treated with caution. Accordingly, 10-mm bars will be considered for Cases (a), (b) and (c) below.

Check movement joint spacing as required in Table 15.1(d) of the Handbook.

(i) *Case (a)*
Maximum spacing complete joints

$$= 4{\cdot}8\,\mathrm{m} + \frac{\mathrm{w}}{\varepsilon}$$

where

$$\varepsilon = \varepsilon_{cs} + \varepsilon_{te} - (100 \times 10^{-6})$$

which is equal to

$$\frac{\alpha}{2} \times T_1 \qquad \text{(Clause B2, Appendix B of the Code)}$$

$$= \frac{12 \times 10^{-6}}{2} \times 30$$

$$= 0{\cdot}000\,18$$

and maximum crack width $W = 0{\cdot}2\,\mathrm{mm}$. Therefore joint centres

$$= 4{\cdot}8 + \frac{0{\cdot}2}{0{\cdot}000\,18 \times 10^3} = 4{\cdot}8 + 1{\cdot}11 = 5{\cdot}9\,\mathrm{m}$$

(ii) *Case (b)*
Alternate partial and complete joints:

$$\text{centres} \ngtr \tfrac{1}{2}S_{max} + 2{\cdot}4\,\mathrm{m} + \frac{W}{\varepsilon}$$

$$S_{max} = \frac{f_{ct}}{f_b} \times \frac{\phi}{2p}$$

$$= \frac{4}{5} \times \frac{10}{2 \times 0{\cdot}0018}\,\mathrm{mm}$$

$$= 2{\cdot}22\,\mathrm{m}$$

Table 3.2

Option	*Joint requirements*	*Minimum area (mm²/m) of deformed high yield bars of diameter*		
		8 mm	*10 mm*	*12 mm*
1 (of the Code)	Continuous construction	1 080	1 350	—
2 (of the Code)	Complete contraction joint at centre line of pool	653	810	968
3 (of Table 15.1 (d) from the Handbook)	(a) Complete joints at 5·9 m c/c	—	405	—
	(b) Alternate partial and complete joints at 4·6 m c/c	—	405	—
	(c) Partial joints at 3·3 m c/c	—	405	—

Therefore joint spacing

$$= (\tfrac{1}{2} \times 2{\cdot}22) + 2{\cdot}4 + 1{\cdot}11 = 4{\cdot}6\,\text{m}$$

(iii) *Case* (*c*)
Partial joints:

$$\text{centres} \not> S_{max} + \frac{W}{\varepsilon}$$

$$= 2{\cdot}22 + 1{\cdot}11 = 3{\cdot}3\,\text{m}$$

A summary of alternative joint spacings and steel requirements is given in Table 3.2.

DESIGN OF POOL FLOOR

3.22. *Floor Slab*

Deduction of the heel dimension (2 m, Section 3.11) leaves a ground slab of size 21 m by 4·4 m. It will be convenient to cast this slab in two bays each

10·5 m length and its width is just within the maximum convenient for hand tamping. A complete movement joint will therefore be introduced in the centre and at each end and side.

Reference to Table 16.1 in the Handbook gives the following details for a 225-mm-thick slab on a sliding layer:

Area of steel = 344 mm^2/m (by interpolation for 200-mm and 300-mm slabs)

Hence 10-mm high yield bars at 225 mm c/c in both directions will suffice. Note that only one layer of reinforcement is used and this is placed with 45-mm cover in the top of the slab.

Y10 @ 225 mm c/c in both directions—top only

3.23. *Horizontal Tension due to Water Pressure*

The horizontal tension due to water pressure is frequently neglected in pools of this type. However, the effect should be checked and may be significant particularly in deeper pools.

For the continuous design without movement joints, tie bars could connect the heel to a more heavily reinforced floor slab. The direct tension may then be transmitted through reinforcement provided in the floor slab.

Alternatively, and for designs in which the presence of movement joints prevents the use of tie bars, the walls should be checked for sliding, assuming a conservative value for the coefficient of friction between base and foundation stratum.

Bibliography

Deacon, R. C. *Watertight Concrete Construction*, Code 46.504, Cement and Concrete Association, London, 1978, 31 pp.

Perkins, P. H. *Swimming Pools*, Second Edition. Applied Science Publishers, London, 1978, 398 pp.

Tomlinson, M. J. *Foundation Design and Construction*, Third Edition. Pitman Publishing, London, 1975, 785 pp.

CHAPTER 4

PRELIMINARY DESIGN OF A CIRCULAR PRESTRESSED CONCRETE TANK

DESIGN INFORMATION

4.1. *Design Codes*

BS 5337: 1976 (referred to as 'the Code').
CP 110: 1972.

4.2. *Other Design Data*

Handbook to BS 5337: 1976 (referred to as 'the Handbook').

4.3. *Loading Conditions*

Water—density equivalent to 9·8 kN/m^3.
Reinforced concrete—density equivalent to 25 kN/m^3.
Soil backfill—density equivalent to 20 kN/m^3.
Soil pressure (active), $k_a = \tan^2 (45 - \phi/2)$.
Sludge—specific gravity 1·02.

4.4. *Client's Requirements*

The concrete tank is to have an internal diameter of 24·0 m and internal height of 10·0 m. The tank is to retain sludge which can have a maximum temperature of 35 °C. A minimum air temperature of −15 °C can occur.

The tank is to be backfilled externally to half its height with granular material and a minimum temperature of 0 °C is to be allowed for this retained material. The tank must be capable of withstanding water pressure before it is backfilled.

4.5. *Proposed Design Details*

It is proposed to have a free sliding joint at the bottom of the tank. The tank will not be stressed until 28 days after casting so that the transfer strength may be taken to be the same as the characteristic strength. The minimum strength for prestressed concrete is 40 N/mm^2 but 50 N/mm^2 will be used for this design.

4.6. *Design Principles*

Design will be based on the limit state requirements of the Code as given by Section 11. The design will also be in accordance with Section 10 of the Code which gives the general requirements for prestressed concrete tanks including the following:

(i) When the tank is full there should be a compression in the concrete in a circumferential direction of at least 1 N/mm^2 after all losses.

(ii) When the tank is empty the tensile stress in the concrete should not be greater than 1 N/mm^2. (However, if the tank is frequently filled and emptied or is empty for prolonged periods it is desirable to avoid tensile stresses.)

(iii) When the foot of the wall is assumed 'free to slide' a bending moment in the vertical direction should be allowed for on the basis of a restraint equal to one-half of that provided by a pinned foot.

(iv) An allowance should be made for the vertical moment caused by circumferential prestressing. The maximum value to be taken for the flexural stress due to this is 0·3 times the compressive stress due to prestress.

It is proposed to prestress the tank circumferentially and reinforce vertically. The minimum practical thickness for a wall of the proposed dimensions is about 250 mm and this thickness will be tried initially in the design.

4.7. *Other Parts of the Structure*

In practice the tank would be covered with a roof which would bear on the wall. However, for the purpose of this design example forces from the roof will be ignored in the design of the tank.

The foundations will also not be considered here but it may be assumed that these consist of piles driven into a firm stratum and carrying a rigid pile cap.

DESIGN CALCULATIONS

4.8. *Ring Stresses due to Retained Liquid*

Referring to Appendix 3 of the Handbook to obtain ring tension, Section A3.2.1, assuming the wall is free to slide at the top and base then

$$f_t = W_g h r$$

where f_t = ring tension per unit length,
W_g = unit weight of liquid,
h = depth of liquid in tank,
r = radius of tank.

$$\text{Maximum } f_t = (9{\cdot}8 \times 1{\cdot}02) \times 10 \times \frac{24}{2}$$

$$= 1201 \text{ kN per metre height}$$

A residual compression of $1{\cdot}0\ \text{N/mm}^2$ is required so the total maximum ring compression required for the forces due to the retained liquid

$$= 1201 + \frac{1{\cdot}0 \times 250 \times 1000}{1000}$$

$$= 1451 \text{ kN per metre height}$$

Ring tension at mid-height (i.e. where backfill terminates)

$$= (9{\cdot}8 \times 1{\cdot}02) \times 5 \times \frac{24}{2}$$

$$= 601 \text{ kN per metre height}$$

Therefore the ring compression required at mid-height

$$= 601 + \frac{1{\cdot}0 \times 250 \times 1000}{1000}$$

$$= 851 \text{ kN per metre height}$$

4.9. *Temperature Stresses*

Stresses act vertically and circumferentially. Two locations will be examined, at the base and at mid-height. At the base the maximum temperature gradient through the walls is +35 °C to 0 °C. At mid-height where the backfill terminates the differential is from +35 °C to −15 °C, i.e. a total of 50 °C. For the calculation of the stresses due to temperature effects the authors are indebted to Chapter 9 of the text book by L. R. Creasy for which the reference is given in the bibliography at the end of this chapter. It should, however, be noted that such calculations involve a number of assumptions; they should be regarded as giving a useful guide rather than exact figures. Designers should therefore assess the importance of this factor in relation to the overall design and may choose to use other methods described in the specialist literature on this topic. The reference given above is expressed in imperial units for which the SI metric equivalents have been calculated and are given below.

The tensile stress on the cold face of the tank (and the equivalent compressive stress on the hot face) due to temperature gradient T°C is given by

$$f_T = 0{\cdot}25\alpha E_c T \text{ N/mm}^2 \text{ (assuming 50\% plastic redistribution)}$$

If

$$\alpha = 13 \times 10^{-6} \text{ per °C}$$

and

$$E_c = 31 \text{ kN/mm}^2$$

then

$$f_T = 101 \times 10^{-3} T$$

Transfer of heat through the wall is influenced by the capacity of the surface to absorb and reject heat to the surrounding air. Since heat is dissipated from the face of the wall to the surrounding air there will be a temperature difference between the air and the adjacent surface. The temperature gradient across the wall is therefore less than the difference in temperature between the air and water. Internally, when the tank is full, the wall surface and the liquid will be at the same temperature.

The temperature gradient T is calculated from

$$T = (T_o - T_i)RF$$

where T_o and T_i are the external and internal air/water temperatures and RF is a reduction factor which may be taken as

$$RF \text{ (air both faces)} = R_c \times U$$

and

$$RF \text{ (air/water)} = \tfrac{1}{2}[1 + (R_c \times U)]$$

where R_c = thickness (m) × thermal resistivity (the thermal resistivity is taken as $1{\cdot}05\,\text{m}^2\,\text{s}\,°\text{C/J m}$ for most normal weight concrete) and U = heat transfer coefficient which varies from about $2{\cdot}6\,\text{J/m}^2\,\text{s}\,°\text{C}$ for 200 mm thickness to $1{\cdot}6\,\text{J/m}^2\,\text{s}\,°\text{C}$ for 400 mm thickness of normal weight concrete. Hence for 0·25 m thickness (say $U = 2{\cdot}3$)

$$RF \text{ (air/water)} = \tfrac{1}{2}[1 + (0{\cdot}25 \times 1{\cdot}05 \times 2{\cdot}3)] = 0{\cdot}8$$

At the tank base

$$T = 35 \times 0{\cdot}8 = 28\,°C$$

and at mid-height

$$T = (15 + 35) \times 0{\cdot}8 = 40\,°C$$

Hence at the tank base

$$f_T = \pm 101 \times 10^{-3} \times 28 = \pm 2{\cdot}9\ \text{N/mm}^2$$

and at mid-height

$$f_T = \pm 101 \times 10^{-3} \times 40 = \pm 4{\cdot}1\ \text{N/mm}^2$$

4.10. *Total Ring Stresses*

Total compressive circumferential force required at the base

$$= 1451 + \frac{2{\cdot}9 \times 250 \times 1000}{1000} = 2176\,\text{kN/m}$$

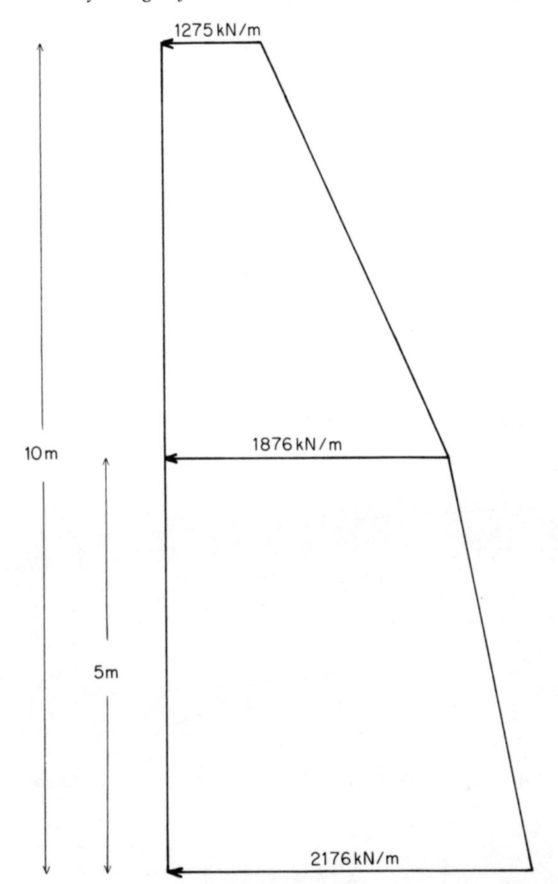

Fig. 4.1. *Ring compression in the tank.*

and at mid-height

$$= 851 + \frac{4{\cdot}1 \times 250 \times 1000}{1000} = 1876\,\text{kN/m}$$

At the top of the wall, ring compression required due to temperature and residual compression requirements is

$$\frac{4{\cdot}1 \times 250 \times 1000}{1000} + \frac{1{\cdot}0 \times 250 \times 1000}{1000} = 1275\,\text{kN/m}$$

The distribution of ring compression is therefore as shown in Fig. 4.1.

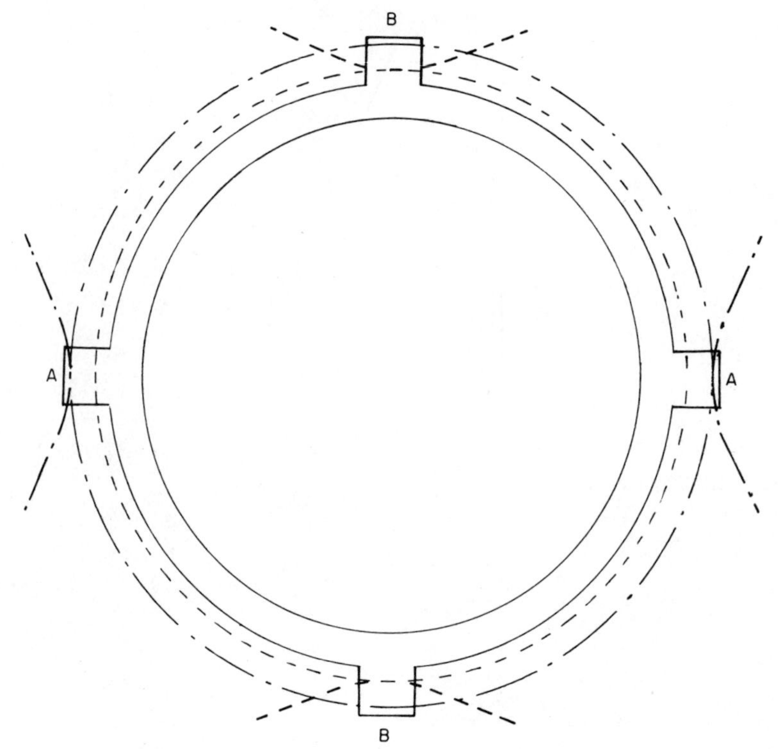

Fig. 4.2. *Jacking points staggered for alternate cables.*

4.11. *Prestressing Forces*

It is proposed to provide four anchorage pilasters at the ends of two diameters at right angles. Cables will be taken round half the perimeter and will be stressed from *both* ends. Each complete perimeter of the tank is to be stressed simultaneously, i.e. four jacks will be required during construction. Jacking points for alternate cables in the height of the structure will be staggered, as shown in Fig. 4.2.

With such a stressing sequence the maximum friction loss occurs at a point mid-way between pilasters.

If x is distance from jacking end to the point of minimum stress

$$x = \frac{\pi \times 24{\cdot}25}{8} = 9{\cdot}5 \text{ m}$$

Referring to CP 110, Clause 4.8.3, and assuming the constant $K = 33 \times 10^{-4}$ and $\mu = 0{\cdot}25$, from Clause 4.8.3.5

$$r_{ps} = \frac{24{\cdot}25}{2} = 12{\cdot}12\,\text{m}$$

$$\left(Kx + \frac{\mu x}{r_{ps}}\right) = \left(0{\cdot}0033 + \frac{0{\cdot}25}{12{\cdot}12}\right) \times 9{\cdot}5$$
$$= 0{\cdot}227$$

$$P_x = P_o e^{-0{\cdot}227} \text{(CP 110, equations 49 and 50)}$$
$$= 0{\cdot}80 P_o$$

Use 15·2-mm low relaxation strand, section area 138·7 mm², characteristic strength 227 kN and a jacking force of 70 % of this value. The relaxation loss given by experimental evidence will be supplied by the manufacturer but 1·1 % is a typical value.

$$P_x = 0{\cdot}8 \times 0{\cdot}7 \times 227 = 127\,\text{kN per strand}$$

4.12. *Prestress Losses*

At the bottom of the tank at transfer

(a) Relaxation of steel taken as 1·1 % (CP110, Clause 4.8.2.2):

$$\text{loss} = 0{\cdot}011 \times 127 = 1{\cdot}4\,\text{kN}$$

(b) Shrinkage—for transfer not earlier than 28 days after casting assume 150×10^{-6} per unit length (CP 110, Table 41, but value reduced by 25 % for partially saturated conditions and stressing at 28 days):

$$\text{loss} = 150 \times 10^{-6} \times 200 \times 138{\cdot}7 = 4{\cdot}2\,\text{kN}$$

(c) Creep of concrete:
Maximum compressive stress when the tank is empty

$$= \frac{2176 \times 10^3}{250 \times 10^3} = 8{\cdot}7\,\text{N/mm}^2$$

Taking creep per unit length as 36×10^{-6} per N/mm² (CP 110, Clause 4.8.2.5) and assuming 15 % losses

$$\text{loss} = 36 \times 10^{-6} \times \frac{8{\cdot}7}{0{\cdot}85} \times 200 \times 138{\cdot}7 = 10{\cdot}2\,\text{kN}$$

(d) Elastic deformation (CP 110, Clause 4.8.2.3):

$$\text{loss} = \frac{1}{2} \times \frac{200}{31} \times \frac{8{\cdot}7}{0{\cdot}85} \times 138{\cdot}7 \times 10^{-3} = 4{\cdot}6\,\text{kN}$$

$$\text{Total loss} = 1{\cdot}4 + 4{\cdot}2 + 10{\cdot}2 + 4{\cdot}6 = 20{\cdot}4\,\text{kN}$$

$$= \frac{20{\cdot}4}{127} \times 100 = 16{\cdot}1\,\%$$

This is slightly higher than the 15% assumed earlier. However, the tank will not normally be empty and the stresses in (c) and (d) will be reduced when the tank is full. The assumed value of 15% loss will therefore be taken without further adjustment.

4.13. *Prestressing Steel*

Force in 15·2-mm strand mid-way between pilasters

$$= 227 \times 0{\cdot}7 \times 0{\cdot}8 \times 0{\cdot}85 = 108\,\text{kN}$$

At the bottom of the tank a prestress force of 2176 kN/m is required (Section 4.10).

$$\text{Number of strands} = \frac{2176}{108} = 20 \text{ per metre height}$$

At mid-height this reduces to

$$\frac{1876}{108} = 17{\cdot}4 \text{ (say 17 per metre height)}$$

At the top this reduces to

$$\frac{1275}{108} = 11{\cdot}8 \text{ (say 12 per metre height)}$$

In practice the intermediate heights would be divided into a series of equal bands superimposed on the diagram of Fig. 4.2. The appropriate number of strands would then be calculated for each band. However, the precise arrangement would be subject to adjustment depending on the stressing system used. If external strands wound on to the outside face are to be used then these would be spaced in accordance with the numbers required per metre calculated as above. They would, of course, require subsequent protection in accordance with CP 110, Clauses 4.9.3 and 8.8.3. The additional concrete is often applied as 'gunite' or 'sprayed concrete', in which case BS 5337, Clause 28, applies. The authors would not, however, recommend the use of hydrated lime in this cover concrete and this is further discussed in Part 2.

4.14. *Maximum Ring Compression*

Maximum soil pressure on wall from backfill (assume $\phi = 35°$ for granular backfill, 20 kN/m^3 density).

$$k_a = \tan^2\left(45 - \frac{\phi}{2}\right) = 0{\cdot}27$$

The ring compression causes

$$f_c = (0{\cdot}27 \times 20) \times 5 \times \frac{24{\cdot}25}{2}$$

$$= 328\,\text{kN per metre height}$$

$$= \frac{328 \times 10^3}{250 \times 1000}$$

$$= 1{\cdot}3\,\text{N/mm}^2$$

The maximum initial stress induced at transfer of prestress will be

$$0{\cdot}9 \times \frac{8{\cdot}7}{0{\cdot}8 \times 0{\cdot}85} = 11{\cdot}5\,\text{N/mm}^2$$

if it is assumed that the *minimum* friction loss is 10% due to the staggered cable arrangement shown in Fig. 4.2.

For uniform prestress CP 110 permits (Table 36) $0{\cdot}4f_{cu}$, i.e. 20 N/mm^2 for Grade 50 concrete.

Transfer stress satisfactory

The prestressing force required was calculated in Section 4.8 for a full tank. For obvious practical reasons the prestressing force is actually applied when the tank is empty. As the tank is filled, however, its circumference increases due to the radial pressure from the contained liquid. This results in increased stress in the prestressing steel and a corresponding decrease in the concrete stress. The resultant service stresses must now be checked.

Assume that when the tank is full the circumference has increased by x m. After losses the force in each strand was found to be 108 kN (Section 4.13) on a section area of $138{\cdot}7\,\mathrm{mm}^2$. In the empty condition the steel stress

$$= \frac{108 \times 10^3}{138{\cdot}7} = 779\,\mathrm{N/mm^2}$$

The concrete stress in the empty condition was $8{\cdot}7\,\mathrm{N/mm^2}$ at the base of the tank (Section 4.12(c)).

Forces may now be calculated for a 1-m-height section at the base of the tank when full.

Closing forces (full)

$$\text{Tension in steel} = 20 \times 138{\cdot}7\left(779 + \frac{xE_s}{2\pi r}\right)\mathrm{N}$$

$$\text{Tension due to backfill} = 328 \times 10^3\,\mathrm{N}$$

Opening forces (full)

$$\text{Concrete compressive force} = (250 \times 1000)\left(8{\cdot}7 - \frac{xE_c}{2\pi r}\right)\mathrm{N}$$

$$\text{Ring tension due to liquid (from Section 4.8)} = 1201 \times 10^3\,\mathrm{N}$$

The backfill will be ignored in checking the maximum service stress in the steel and in checking for concrete tension. Hence, for equilibrium

$$20 \times 138{\cdot}7\left(779 + \frac{xE_s}{2\pi r}\right) = \left[(250 \times 10^3)\left(8{\cdot}7 - \frac{xE_c}{2\pi r}\right)\right] + (1201 \times 10^3)$$

Therefore

$$\frac{xE_c}{2\pi r}\left(250 + 2{\cdot}774\frac{E_s}{E_c}\right) = 2175 + 1201 - 2161 = 1215$$

Putting $E_s = 200\,\mathrm{kN/mm^2}$ and $E_c = 31\,\mathrm{kN/mm^3}$

$$\frac{E_s}{E_c} = 6{\cdot}45$$

$$\frac{xE_c}{2\pi r} = \frac{1215}{250 + (2{\cdot}774 \times 6{\cdot}45)} = 4{\cdot}5\,\mathrm{N/mm^2}$$

Hence concrete stress when full

$$= 8{\cdot}7 - 4{\cdot}5 = 4{\cdot}2\,\mathrm{N/mm^2}$$

The temperature stress may reduce this value on the external face by 2·9 N/mm^2 (Section 4.9) giving a residual stress of 1·3 N/mm^2 which exceeds the minimum value of 1 N/mm^2 required.

Minimum service concrete compressive stress satisfactory

The corresponding tension in the steel

$$= 779 + (4{\cdot}5 \times 6{\cdot}45) = 808\ \text{N/mm}^2$$

i.e.

$$\frac{808 \times 138{\cdot}7}{10^3} = 112\,\text{kN per strand}$$

or,

$$\frac{112}{227} \times \frac{100}{0{\cdot}87} = 57\,\%\ \text{of the design ultimate strength}$$

Maximum service steel tension satisfactory

The maximum service compressive stress occurs in the concrete when the tank is empty and the backfill is in place. Emptying the tank and placing the backfill both cause contraction of the tank circumference from the full condition just calculated. Assuming a contraction of y m the forces on the empty tank will be as follows:

Closing forces (empty)

$$\text{Tension in steel} = 20 \times 138{\cdot}7\left(808 - \frac{yE_s}{2\pi r}\right)\text{N}$$

$$\text{Tension due to backfill} = 328 \times 10^3\ \text{N}$$

Opening forces (empty)

$$\text{Concrete compressive force} = (250 \times 1000)\left(4{\cdot}2 + \frac{yE_c}{2\pi r}\right)\text{N}$$

Therefore for equilibrium

$$\frac{yE_c}{2\pi r}\left(250 + 2{\cdot}774\frac{E_s}{E_c}\right) = 2241 + 328 - 1050 = 1519$$

$$\frac{yE_c}{2\pi r} = \frac{1519}{250 + (2{\cdot}774 \times 6{\cdot}45)} = 5{\cdot}7\ \text{N/mm}^2$$

The maximum service direct compressive stress in the concrete

$$= 4{\cdot}2 + 5{\cdot}7 = 9{\cdot}9\ \text{N/mm}^2$$

If the tank is emptied when maximum temperature stresses are present, a further compressive stress of 2·9 N/mm^2 is added on the inside face to give 12·8 N/mm^2. These stresses compare with a maximum permitted direct stress of 12·5 N/mm^2 and maximum permitted bending stress of 16·7 N/mm^2 (CP 110, Table 32).

Maximum service concrete compressive stress—satisfactory

Ultimate conditions will be checked by considering the effect of liquid pressure producing a ring tension of 1·6 × 1201 kN/m at the base of the tank.

The steel stress for the empty tank condition without reduction for backfill pressure will be taken, i.e. 779 N/mm^2 as before, with the corresponding concrete stress of 8·7 N/mm^2. Hence, under ultimate conditions, if the circumference increases by z m

Closing forces (*ultimate*)

$$\text{Tension in steel} = 20 \times 138{\cdot}7\left(779 + \frac{zE_s}{2\pi r}\right)\text{N}$$

Opening forces (*ultimate*)

$$\text{Concrete compressive force} = 250 \times 1000\left(8{\cdot}7 - \frac{zE_c}{2\pi r}\right)\text{N}$$

$$\text{Ring tension due to liquid} = 1{\cdot}6 \times 1201 \times 10^3 = 1922 \times 10^3\,\text{N}$$

For equilibrium

$$20 \times 138{\cdot}7\left(779 + \frac{zE_s}{2\pi r}\right) = 250 \times 1000\left(8{\cdot}7 - \frac{zE_c}{2\pi r}\right) + (1922 \times 10^3)$$

Therefore

$$\frac{zE_c}{2\pi r}\left(250 + 2{\cdot}774\frac{E_s}{E_c}\right) = 2175 + 1922 - 2161 = 1936$$

and

$$\frac{zE_c}{2\pi r} = \frac{1936}{250 + (2{\cdot}774 \times 6{\cdot}45)} = 7{\cdot}2\,\text{N/mm}^2$$

Hence concrete stress

$$= 8{\cdot}7 - 7{\cdot}2 = 1{\cdot}5\,\text{N/mm}^2$$

and steel stress

$$= 779 + (7{\cdot}2 \times 6{\cdot}45) = 826\,\text{N/mm}^2$$

or

$$\frac{826 \times 138{\cdot}7}{10^3} = 115\,\text{kN}$$

which is considerably less than the ultimate design values of $0{\cdot}87 \times 227 = 197\,\text{kN}$.

Ultimate conditions satisfactory

4.15. *Vertical Bending Moment due to Contained Liquid*

Considerations affecting reinforcement for vertical section are

(i) bending moment caused by circumferential prestress including the temporary condition when stressing, Section 4.6(iv);
(ii) tensile stress caused by temperature differential through section;
(iii) tensile stress due to bending moment in vertical plane from retained sludge;
(iv) bending moment due to backfill.

Bending moments for a sliding base and pinned base may be obtained from Appendix A3.2.2 of the Handbook. Paragraph (a) of this Appendix states

> if the tank is filled when the wall base is free to slide, no moment is induced in the vertical direction

However, the pinned condition must also be considered for the reasons given in Section 4.6 (iii).

Paragraph (b) for walls pinned at the base gives the following equation for the moment at any height:

$$M = \text{coefficient} \times (W_g h^3 + qh^2)$$

In this instance $q = 0$ and referring to Table A3.7 of the Handbook

$$\frac{h^2}{dt} = \frac{10^2}{24{\cdot}25 \times 0{\cdot}25} = 16{\cdot}5$$

This slightly exceeds the values in the tables to the Handbook. Since higher values may occur in practical designs additional tables are included in the appendix to Chapter 1 of this book. They are drawn from the same reference as those in the Handbook and are reproduced with the kind permission of the Portland Cement Association. The full reference is given in the bibliography at the end of this chapter. Hence the following coefficient values are obtained and the values of $W_g h^3$ and M calculated:

From the Handbook, Table A3.7, for $\dfrac{h^2}{dt} = 16$

$$\text{Maximum coefficient} = +0{\cdot}0029 \text{ at } 0{\cdot}9\,h$$

From Table A3.7a of this book (p. 17) for $\dfrac{h^2}{dt} = 20$

$$\text{Maximum coefficient} = 0{\cdot}0024 \text{ at } 0{\cdot}9h$$

Take 0·0029:

$$M = 0{\cdot}0029(9{\cdot}8 \times 1{\cdot}02) \times 10^3 = 29\,\text{kN m/m}$$

It should be noted that

(a) these values indicate tension on the outside face,
(b) 'h' is measured from the top of the tank and $1{\cdot}0h$ is therefore the bottom,
(c) values to be halved for a free sliding base.

Take maximum bending moment in vertical direction due to the semi-pinned condition

$$= \frac{29}{2} = 14{\cdot}5\,\text{kN m/m}$$

4.16. *Vertical Bending due to Temperature*

The temperature stresses in the lower half were 2·9 N/mm² tension on the external face and corresponding compression internally. Hence

$$M_T = 2{\cdot}9 \times \frac{bd^2}{6}$$
$$= \frac{2{\cdot}9 \times 1000 \times 250^2}{6} \times 10^{-6}$$
$$= 30{\cdot}2\,\text{kN m/m}$$

4.17. *Vertical Reinforcement (Outer Face)*

Class A exposure will be taken (see the Handbook, paragraph 10.1).

From Sections 4.15 and 4.16, the maximum total moment giving tension on the external face

$$= 14{\cdot}5 + 30{\cdot}2 = 44{\cdot}7\,\text{kN m/m}$$

The Handbook, Table A1.1.2, gives 25-mm bars at 200-mm spacings (with $f_{st} = 115\,\text{N/mm}^2$).

*Y*25 @ 200 *c/c*

4.18. *Vertical Bending due to Earth Pressure*

With tank empty and backfilled to mid-height consider a tank 5 m high:

$$\frac{h^2}{dt} = \frac{5^2}{24{\cdot}25 \times 0{\cdot}25} = 4{\cdot}1$$

Soil pressure (assuming adequate drainage) is equivalent to external liquid of density $0{\cdot}27 \times 20 = 5{\cdot}4\,\text{kN/m}^3$ (Section 4.14). Maximum coefficient from Table A3.7 of the Handbook (see Section 4.15) is now 0·0118 at $0{\cdot}8 \times 5 = 4$ m below mid-height or again 1 m above base. Hence

$$M = 0{\cdot}0118 \times 5{\cdot}4 \times 5^3 = 8{\cdot}0\,\text{kN m/m}$$

(causing tension on internal face). Take one-half as before = 4 kN m/m.

4.19. *Vertical Bending due to Prestressing*

At the base of the tank the applied circumferential prestress force was 2176 kN/m after losses (see Section 4.10). This was composed of a uniform load of 1275 kN/m and an approximately triangular load varying from zero at the top of the tank to 2176 − 1275 = 901 kN/m at the base.

From the formulae given in Appendix A3.2.1 of the Handbook

$$1275 = q \times \frac{24{\cdot}5}{2}$$

Therefore

$$q = 104\,\text{kN/m}^2$$

and

$$901 = W_g \times 10 \times \frac{24{\cdot}5}{2}$$

$$W_g = 7{\cdot}4\,\text{kN/m}^3$$

Hence the prestressing produces radial pressures equivalent (but of opposite sign) to those induced by a gas pressure of $104\,\text{kN/m}^2$ acting in combination with a contained liquid of unit weight $7{\cdot}4\,\text{kN/m}^3$. Referring to the Handbook, Table A3.7 gives a maximum coefficient of 0·0029 at 1 m above the base. Hence the vertical bending moment

$$= \text{coefficient} \times (W_g h^3 + qh^2)$$

$$= 0{\cdot}0029 \times [(7{\cdot}4 \times 10^3) + (104 \times 10^2)]$$

$$= 51{\cdot}6\,\text{kN m/m}$$

Taking one-half as before

$$M = 25{\cdot}8\,\text{kN m/m}$$

4.20. *Vertical Bending Moment due to Backfill and Prestress (Base)*

From Sections 4.18 and 4.19, the maximum moment due to the backfill pressure and prestress giving tension on the external face at the tank base when empty

$$= 4 + 25{\cdot}8 = 29{\cdot}8\,\text{kN m/m}$$

4.21. *Vertical Bending Moment at Transfer (Base)*

Check also this moment at transfer assuming an immediate elastic loss (see Section 4.12) of

$$\frac{4{\cdot}6}{127} \times 100 = 3{\cdot}6\%$$

The ring prestress at the base (Section 4.10) was 2176 kN/m. Taking a minimum friction loss of 10% as in Section 4.14 and immediate elastic loss as above, initial ring compression at base

$$= \frac{2176}{0{\cdot}8} \times \frac{0{\cdot}9}{0{\cdot}85} \times (1 - 0{\cdot}036) = 2776\,\text{kN/m}$$

Similarly at the top, initial ring compression

$$= \frac{1275}{2176} \times 2776 = 1627\,\text{kN/m}$$

The corresponding radial forces calculated as in Section 4.19 are given by

$$1627 = q \times \frac{24{\cdot}5}{2}$$

$$q = 133\,\text{kN/m}^2$$

and

$$2776 - 1627 = W_g \times 10 \times \frac{24{\cdot}5}{2}$$

$$W_g = 9{\cdot}4\,\text{kN/m}^3$$

The vertical bending moment

$$= 0{\cdot}0029 \times \{(9{\cdot}4 \times 10^3) + (133 \times 10^2)\} = 65{\cdot}8\,\text{kN m/m}$$

Since at this stage the backfill is not placed the total moment is simply one-half this value, i.e. 32·9 kN m/m.

4.22. *Vertical Reinforcement (Inner Face)*

The vertical bending moment at transfer due to prestress (Section 4.21) exceeds that in service (Section 4.20) and the value of 32·9 kN m/m will therefore be taken. However, to allow for the temporary effects of solar gain during construction (i.e. before backfilling) this will be increased to 44·7 kN m/m as used for the outer face and the same reinforcement is therefore required.

Y25 @ 200 c/c

4.23. *Vertical Bending Moment at Mid-height*

By inspection of the Handbook, Table A3.7, the moments at mid-height due to the contained liquid when full or due to ring prestress when empty are negligible. However, the temperature stress at this height (Section 4.9) is increased to $4{\cdot}1\ \text{N/mm}^2$ and (referring to Section 4.16) the vertical moment due to this effect

$$= 30{\cdot}2 \times \frac{4{\cdot}1}{2{\cdot}9} = 42{\cdot}7\,\text{kN m/m}$$

giving tension on the external face.

This is practically equivalent to the bending moment provided for at the base (Section 4.17) and the external reinforcement previously calculated for that position will therefore be carried through to the top of the tank.

Outer face Y25 @ 200 c/c throughout

When the tank is empty solar effects above the backfilled mid-height position will result in tension on the inner face. Recent research (see paper by M. J. N. Priestley in the bibliography at the end of this chapter) has shown that significant temperature gradients can occur due to the effect of incident solar radiation. For preliminary design purposes it will be assumed that these effects may be of the same order as the above calculated mid-height temperature moment of the opposite sign due to hot liquid on the internal face. Hence the reinforcement calculated for the internal face at the base will also be carried through to the top of the tank.

Inner face Y25 @ 200 c/c throughout

4.24. *Vertical Bending Moment during Prestressing*

If the prestress is applied by 'winding' external tendons then the effect is to 'squeeze' the tank progressively from the base upwards, thereby inducing a temporary vertical moment. Where the prestress is applied (as assumed in this example) by individual tendons anchored at pilasters it will often be impractical to apply it uniformly over the height of the tank.

Applying the Code rule noted in Section 4.6 (iv) the flexural stress due to this cause at transfer, taking immediate losses as in Section 4.21

$$= 0{\cdot}3 \times 11{\cdot}5 \times 0{\cdot}9 \times 0{\cdot}964 = 3{\cdot}0\,\text{N/mm}^2$$

Hence vertical bending moment

$$= 3{\cdot}0 \times \frac{1000 \times 250^2}{6} \times 10^{-6} = 31{\cdot}2\,\text{kN m/m}$$

The vertical reinforcement provided caters for a total vertical bending moment of 44·3 kN m/m. Hence an allowance of

$$44{\cdot}3 - 31{\cdot}2 = 13{\cdot}1 \text{ kN m/m}$$

is available to resist moments due to temperature gradient which may exist at the time of stressing and in most circumstances this would be reasonable.

Vertical M during stressing satisfactory

4.25. *Minimum Reinforcement*

Theoretically there is no vertical or horizontal restraint during the curing period. However, even if base restraint is very small some tension must develop due to temperature and shrinkage effects. Hence provide 0·30% minimum reinforcement in accordance with the Code, Clause 4.11.2:

$$A_s = \frac{0{\cdot}3}{100} \times 250 \times 1000 = 750 \text{ mm}^2\text{/m}$$

The amount already provided in the vertical direction exceeds this but horizontal reinforcement is required to control secondary stresses prior to transfer.

Horizontal Y12 @ 300 c/c both faces

4.26. *Final Notes*

The calculations illustrated in this chapter are those for preliminary design. For detailed design the reader is referred to the bibliography which follows. In particular it should be noted that certain secondary effects have been ignored at this stage but should be included in a final design.

One aspect not considered in the literature and often ignored by designers is the effect of grip 'pull in' at the anchorages. As this occurs, friction on the section of tendon near the anchorage is reversed. The result is that the loss due to grip 'pull in' may be confined to a part of the tendon close to the anchorage. Dependent on the amount of pull in and other factors, this can result in a loss of, say, 20% near the anchorage.

With jacking points at alternate pilasters, as in Fig. 4.2, each tendon has a length equal to half the circumference. The maximum friction loss in the individual tendon therefore occurs at its mid-point.

Referring to Section 4.11 and calculating the friction loss at this tendon mid-point

$$x = 19{\cdot}0\,\text{m}$$

and

$$Kx + \frac{\mu x}{r_{ps}} = 2 \times 0{\cdot}227 = 0{\cdot}454$$

Hence

$$P_x = 0{\cdot}64\,P_o$$

Thus, the average force at a pilaster position (taking an assumed 20% friction loss near the anchorage) becomes

$$P_x = \frac{(0{\cdot}64 + 0{\cdot}80)}{2}\,P_o = 0{\cdot}72\,P_o$$

which is less than the value $0{\cdot}8P_o$ taken previously.

Clearly the result is highly dependent on the amount of 'pull-in' and the length over which reverse friction occurs.

The reader will note the major contribution made by temperature effects to the calculated moments and stresses in this example. The Code does not give guidance on this aspect of design and reference must be made to the specialist literature. However, the observant designer will see tanks constructed with substantially less allowance for temperature stresses than those made in this example. In the absence of specific Code guidance the designer must make his own assessment of the appropriate allowance for temperature effects and the methods to be used for this purpose in design. It is, however, worth noting the very substantial allowance which is now made in bridge deck designs for stresses due to temperature gradients and which were omitted entirely in UK designs until recently. More directly relevant is a paper by M. Schupak, presented at the ACI symposium listed in the bibliography, which refers to cracking 3–5 years after construction apparently caused by differential temperature and other secondary effects.

It is appropriate to make one further comment on the calculation of a reinforced concrete section in water retaining structures. The method of design included in earlier editions of the Code (CP 2007) was similar to the present Code 'alternative method'. As was seen in Chapter 2, this is somewhat tedious to apply when compared with reinforced concrete designed for other uses. When the new Code was published designers generally found that calculation of the crack width for limit state design was even more tedious than the 'alternative method' retained from the earlier Code. Consequently they either applied this method or, if limit state design was used, it was by reference to the allowable stresses of Table 1 of the Code. The extent to which publication of the Handbook Appendix 1 tables has changed this situation is demonstrated in the vertical design of the tank walls. Quite clearly the use of these tables provides the simplest possible design procedure for liquid retaining reinforced concrete.

Bibliography

American Concrete Institute. *Symposium on Concrete Construction in Aqueous Environments, Seattle* 1962. ACI publication SP 8.

American Concrete Institute Committee 344. Design and construction of circular prestressed concrete structures. Title No. 67–40. *J. ACI*, September 1970.

Creasy, L. R. *Prestressed Concrete Cylindrical Tanks*. John Wiley & Sons, New York, and Contractors Record Ltd, London, 1961, 216 pp.

Portland Cement Association. *Circular Concrete Tanks without Prestressing*. Concrete Information, Illinois, 1941, 32 pp.

Priestley, M. J. N. Ambient thermal stresses in circular concrete tanks. *J. ACI*, *October* 1976, *pp*. 553–60.

PART 2

Specification and Construction

CHAPTER 5

MATERIALS USED IN CONSTRUCTION

5.1. *Introduction*

The range of structural materials which can be used successfully for liquid retaining structures is at the present time limited, and the authors believe this position will continue for the foreseeable future.

This does not mean that there will be no progress or improvement in the materials themselves and in their method of use. On the contrary, research and development is going on continuously in most countries of the world and improved versions of all the existing materials are constantly coming on to the market.

In the case of Portland cement concrete, which is the main material used in the type of structure covered by this book, this development is mainly directed towards the use of fibres dispersed throughout the concrete mix, and the practical use of polymer concrete.

The fibres which now show the most promise are ferrous wire, polypropylene and special alkali-resistant glass. The object of using the fibres is to obtain an increase in impact resistance and in the strength of the concrete. At the time of writing, favourable results are being obtained from ferrous wire, and alkali-resistant glass; high-grade precast pipes complying with the performance tests in BS 556 are being made with these two types of concrete. However, the presence of the fine steel wires close to the surface of the concrete results in brown staining if the fibres are not stainless steel. Although this staining has no effect on the long-term durability of the concrete, it is obviously not acceptable for concrete exposed to view.

There are two types of polymer concrete; one in which the cement is replaced by an organic polymer such as epoxide or polyester resin and latex emulsions, and the other in which a monomer is added to the concrete and then polymerized by heat or gamma radiation. Neither of these concretes are at present a practical substitute for normal concrete as the basic material for liquid retaining structures.

Considerable interest has been shown in recent years in ferro-cement, particularly for boats, and the authors believe there are possibilities for the use of this material for relatively small liquid retaining structures.

Most of the materials used for liquid retaining structures are covered by National Standards in many countries. Generally, British Standard Specifications lay down basic requirements and standard tests for the materials, but leave the details of manufacture to the producer. On the other hand Codes of Practice usually deal with principles of design and methods of construction for the particular work to which they refer.

Appendix 1 gives the titles and numbers of the more important British and United States Standards and Codes of Practice relevant to liquid retaining structures.

The authors feel that notes on the basic properties of the materials in general use for liquid retaining structures will be useful to readers and these are given in this chapter.

Information on manufacture and details of the relevant British Standards are not included but are referred to in appropriate cases.

5.2. *Cements*

For the purpose of this book cements are classified as Portland cements and non-Portland cements.

The quantity of Portland cement used in the construction industry in the UK far exceeds that of all other types and in 1977 amounted to about 17 million tonnes.

Portland Cements

ORDINARY AND RAPID HARDENING PORTLAND CEMENT

In the UK these two cements are covered by BS 12—*Portland Cement—Ordinary and Rapid Hardening*, and form the bulk of the Portland cements used in liquid retaining structures, with ordinary Portland predominating.

The basic difference between the two cements is the rate of gain of strength; the increase is largely due to the finer grinding of the rapid hardening cement which usually has a specific surface of about 4300 cm^2/g. The rapid hardening is accompanied by an increase in the rate of evolution of heat of hydration which in turn raises the temperature of the maturing concrete during the first 15–40 h after casting. It should be noted, however, that the rate of hardening and evolution of heat of hydration does not depend on the fineness of grinding alone, and that the chemical composition of the cement also plays a part.

Sulphate Resisting Portland Cement

The relevant British Standard is BS 4027, and the cement is similar in strength and other physical properties to ordinary Portland cement. It is generally darker in colour than ordinary and rapid hardening Portland cements and the darker shades sometimes approach that of high alumina cement. The essential feature of sulphate resisting Portland cement is the limitation of the tricalcium aluminate (C_3A) content to a maximum of 3 %.

It is the C_3A in Portland cement which reacts with sulphates in solution resulting in the formation of the compound ettringite. The reaction is expansive in character, resulting in an increase in volume of the concrete or mortar and a reduction in strength.

Some sulphate resisting Portland cements have a lower heat of hydration than some ordinary Portland cements and in certain circumstances this can be advantageous in helping to reduce the tendency to thermal contraction cracking.

It is advisable to consult the cement manufacturer before using any type of admixture with sulphate resisting Portland cement, but chloride-containing admixtures should not be used as the sulphate resistance in the long term will be reduced. (See also the latest revision to CP110—*The Structural Use of Concrete*.)

Sulphate resisting Portland cement, in common with all Portland cements, is vulnerable to acid attack. However, where dilute sulphuric acid is involved, there would be some advantage in using sulphate resisting Portland cement instead of ordinary Portland cement.

White and Coloured Portland Cements

White Portland cement is a true Portland cement and complies with BS 12. The special point about this cement is that the raw materials are specially selected; the clay is a white china clay and the manganese and iron content is kept to an absolute minimum.

Coloured Portland cements, other than white and the pastel shades, generally consist of ordinary Portland with a pigment ground in at the works. The pigments used are covered by BS 1014—*Pigments for Cement and Concrete*. Coloured Portland cements are now included in BS 12.

Coloured Portland cement, other than white, cannot now be obtained on the retail market in the UK.

A question is sometimes raised as to whether it is necessary to increase the cement content of a concrete or mortar mix to allow for the presence of the pigment in the mix. It is not possible to give an answer to this in the form of a straight 'yes' or 'no'. Where durability and/or impermeability are important factors, the authors consider that allowance for the pigment should be made by increasing the specified cement content by the weight of pigment. Where strength is overriding, then the only satisfactory solution is to make trial mixes because the strength of ordinary Portland cements varies within certain limits.

Ultra-high Early Strength Portland Cement

This is a cement of the true Portland type which appeared on the UK market some years ago under the trade name of 'Swiftcrete'. At the time of writing there is no British Standard for this cement.

It is described by the manufacturers as an extremely finely ground Portland cement which contains a higher proportion of gypsum than ordinary Portland, but otherwise complies with BS 12; it contains no other additives.

The specific surface of the cement is between 7000 and 9000 cm^2/g compared with an average ordinary Portland of 3500 cm^2/g, and 4300 cm^2/g for an average rapid hardening Portland.

The tests carried out by the Agrément Board confirmed that the strength at 24 h is not less than the strength of a similar mix of concrete made with rapid hardening Portland cement at 7 days. The original Certificate made a number of statements relating to the use of this cement and included a table giving a detailed comparison between 'Swiftcrete' and ordinary and rapid hardening Portland cements.

Non-Portland Type Cements

Supersulphated Cement

Supersulphated cement consists of granulated blastfurnace slag, calcium sulphate and a small proportion of Portland cement or lime.

It is sometimes used for structural concrete in foundations, retaining walls, and other parts of a structure which are in direct contact with soil or ground water containing a high concentration of sulphates and/or some dilute mineral acids.

There are some practical difficulties in its use on site compared with Portland cements; as an example, the rate of hardening is seriously reduced by a fall in temperature. It should not be mixed with any other type of cement, and admixtures should not be used without the agreement of the manufacturers. The relevant British Standard is BS 4248.

This cement is no longer manufactured in the UK but is sometimes imported from the Continent.

High Alumina Cement (HAC)

A few years before this book was written a great controversy arose in the UK over the use of HAC concrete in building structures. This was caused by the unfortunate failure of two roof beams in a school. A considerable amount of ill-informed comment was made in the national press and in some technical journals. Because of this the authors consider it desirable to present a brief and balanced appraisal of HAC concrete and to show how this differs in important respects from Portland cement concrete.

High alumina cement is covered by BS 915. About 60 % of the world's requirement of this cement (outside the USSR) is produced by Lafarge Fondu International; of this, about 90 % is used for refractory concrete where structural strength may be of little significance.

High alumina cement differs fundamentally in chemical composition from Portland cement as it consists predominantly of calcium aluminates. It is a much darker colour (dark grey) than ordinary and rapid hardening Portland cements. The lighter shades of HAC and the darker shades of sulphate resisting Portland cement may approach each other in colour. The cement sets and hardens when mixed with water, and under normal conditions of temperature the setting time is similar to that of ordinary Portland cement. However, with increase in ambient temperature, under the normal climatic conditions existing in the UK, the setting time tends to increase, whereas the reverse is the case with Portland cement. In countries where ambient temperatures are higher it should be noted that with temperatures above about 30 °C, the setting time is reduced and the set becomes increasingly faster and in extreme cases may even 'flash' set. On the other hand, the rate of gain of strength of HAC is very rapid, normally reaching about 80 % of its strength in 24 h, compared with Portland cement which reaches about 80–85 % of its ultimate strength in 28 days. This rapid increase in strength is accompanied by a rapid evolution of the heat of hydration. This has advantages and disadvantages. It is extremely useful when working at low temperatures; also it enables emergency repairs and similar work to be carried out within a short time. Wet curing is essential and the concrete must be placed in relatively thin layers so that the heat can be dissipated quickly. It is usually recommended that the thickness of 'lifts' should be restricted so that no joint is more than 500 mm from a cured surface. This means that a wall up to 1000 mm thick can be cast continuously without horizontal joints, but steel formwork should be used.

High alumina cement concrete which has been correctly proportioned, placed and cured, exhibits improved resistance, compared with Portland cement concrete, to many chemical compounds, including sulphates, sugars, vegetable oils and certain dilute acids including humic acids. High alumina cement is not classified as a chemically resistant cement and chemicals including some chlorides and alkalis, which have little effect on Portland cement concrete, will attack HAC concrete.

To achieve chemical resistance and long-term durability, the same basic principles used for Portland cement concrete should be adopted. These are high cement content—not less than 400 kg/m^3, low water/cement ratio—not higher than 0·40, thorough compaction and careful curing. This gives a dense, impermeable concrete which is essential for durability.

High alumina cement is more coarsely ground than ordinary Portland cement (specific surface of HAC is about 3000 cm^2/g compared with about 3500 cm^2/g for an average ordinary Portland cement). This, together with the physical characteristics of the cement particles, enables workable mixes to be prepared with lower free water/cement ratios than mixes with Portland cement having the same workability.

It is at this point that mention must be made of the most controversial matter relating to HAC, namely, 'conversion'. A great deal has been written about conversion, but briefly, the hydrates which cause HAC to develop strength rapidly are metastable and convert into a denser and more stable form. Unpublished work by D. E. Shirley, and a report by French and others in the journal *Concrete*, August 1971, pp. 3–8, and a report by Dr C. M. George of Lafarge Fondu International in 1975 (see the bibliography at the end of this chapter) suggest that the phenomenon of conversion could in fact be allowed for in a relatively simple manner in the mix design procedure. In short, the effect of conversion depends mainly on the rate at which it takes place, and this in turn is determined largely by the temperature during the maturing period and of the environment in which the concrete exists during its working life. The original water/cement ratio will basically determine the strength of the concrete after conversion. There is evidence to suggest that conversion may be of little practical significance provided the concrete mix was originally designed for full compaction at a water/cement ratio appropriate to the maximum sustained temperature during the life of the structure. An acceptable formula incorporating the water/cement ratio and temperature and relating these to minimum strength has not, unfortunately, been generally agreed. The nearest to this was contained in an article in the *New Civil Engineer* on 21 March 1974, pp. 40 and 43, and the information given in the publication written by Dr C. M. George.

A further point in connection with the controversial conversion is that after the minimum point has been reached, a high-quality HAC concrete made with a low water/cement ratio will start to increase in strength.

The use of HAC in water retaining structures is likely to be confined to

(a) emergency repairs;
(b) the surfacing of concrete with gunite or hand-applied rendering where the chemical resistance of the cement is an advantage compared with Portland cement.

The Building Research Establishment at Garston developed a relatively simple and reliable test to determine on site whether a particular concrete member contains HAC. Details of this test are given in BRE Information Sheet No. IS.15/74, dated October 1974.

It must be mentioned that when there has been some shortcoming in the original mix design, e.g. a too high water/cement ratio, then the effect of conversion can be serious. This can result in a significant reduction in strength and increase in vulnerability to chemical attack, particularly by caustic alkalis. There are examples of caustic alkalis being leached from Portland cement roof screeds due to defective waterproof membranes and penetrating into HAC concrete structural slabs and beams, resulting in serious attack on the HAC.

Chemically Resistant Cements

These special cements are not used for concrete except where very small quantities are required, but are used for mortar for bedding and jointing chemically resistant tiles and bricks. This type of construction is used in lining tanks which hold very aggressive solutions.

The two basic types of cement are resin cements and silicate cements; certain grades of the latter are resistant to high temperatures. Cements can now be obtained which are resistant to most chemicals in common use with the general exception of hydrofluoric acid in concentrations above 40%.

Chemically resistant cements usually consist of a powder and a special gauging liquid (sometimes called a syrup) which are mixed together in the prescribed proportions.

Comparison of UK and US Cements

In view of the employment of consultants and contractors from both the

Table 5.1

Nature of cement	*ASTM No. and Type*	*British Standard No.*
Ordinary Portland	C 150–67, Type I	BS 12: 1958
Sulphate resisting/low heat	C 150–67, Type II	No equivalent material
Rapid hardening	C 150–67, Type III	BS 12: 1958
Low heat	C 150–67, Type IV	BS 1370: 1958
Sulphate resisting	C 150–67, Type V	BS 4027: 1966
Air entraining (various)	C 175–67, Types IA, IIA, IIIA	No equivalent materials

UK and USA on international contracts, it is considered that some general information on Portland type cements from these two countries would be useful.

The direct comparison of Standard Specifications of one country with those of another can be very misleading. The requirements are different, and the methods of test are different. However, Table 5.1 is intended to show approximate comparisons. For example, the most that can be said is that for conditions where a sulphate resisting Portland cement to BS 4027 would be specified, it is likely that a US Portland cement to C 150–67, Type V would also be satisfactory.

5.3. *Aggregates for Concrete*

General Considerations

The type of aggregate normally used for concrete liquid retaining structures comes under the classification of 'aggregates from natural sources', for which the relevant British Standard is BS 882. Aggregates of this type consist of crushed rock, river and pit gravels and sea-dredged shingle. There appears to be no technical reason why some of the artificial lightweight aggregates should not be used for liquid retaining structures, as they have been used with success in structural concrete for buildings and in marine structures.

Tests on aggregates are covered by BS 812—*Methods for Sampling and Testing of Mineral Aggregates, Sands and Fillers*. It is important to note that BS 882 contains the following statement:

> No simple tests for durability and frost resistance of concrete or for corrosion of reinforcement can be applied and experience of the properties of concrete made with the type of aggregates in question and a knowledge of their source are the only reliable means of assessment.

Shrinkage of natural aggregates, when excessive, can create problems. In the UK, this characteristic has so far only appeared in certain aggregates in Scotland, but in other parts of the world it has proved to be a significant factor in the selection of aggregate sources. For detailed information on shrinkable aggregates in the UK, the reader is referred to the Building Research Establishment's Digest No. 35—*Shrinkage of Natural Aggregates in Concrete.*

Alkali–aggregate reaction is another matter which should be considered when dealing with aggregates which are basically siliceous and obtained from an untried source and there is no long-term experience with their use in concrete. This aspect of concrete technology is discussed briefly in Chapter 6.

Matters which can be important, depending on the circumstances of each case, include the following:

(a) Source and classification.
(b) Particle shape and surface texture, including flakiness index.
(c) Grading, including clay, silt and dust content.
(d) Organic impurities (generally fine aggregate only).
(e) Salt content, including chloride and sulphate concentration.
(f) Crushing strength and absorption.
(g) Shell content, generally sea-dredged aggregates only.

It has been found particularly difficult to place limits on organic impurities. The most satisfactory procedure is to carry out tests using the 'suspect' aggregate and a known 'pure' aggregate as control, to ascertain what changes occur in the setting time, rate of gain of strength, and the 7 and 28 day cube strengths between the two groups.

The UK code for concrete water retaining structures (BS 5337) places an upper limit of 3% on absorption of aggregates. The authors have been unable to find any reliable published information which justifies this limitation. It is the quality and quantity of the cement paste which plays the dominant role in determining the impermeability and durability of the concrete and the protection of the reinforcement. This is discussed in detail in Chapter 6. It is of interest to note that CP 110 does not impose any restrictions on absorption and CP 102—*Protection of Buildings Against Water from the Ground* limits absorption to 5%. Lea, in his standard text book *The Chemistry of Cement and Concrete*, mentions a figure of 10% absorption as being a probable upper limit.

Sea-dredged Aggregates

These aggregates are being used to an increasing extent, particularly in the

Table 5.2

Aggregate size	*Maximum shell content—dry weight*
40 mm ($1\frac{1}{2}$ in)	2%
20 mm ($\frac{3}{4}$ in)	5%
10 mm ($\frac{3}{8}$ in)	15%
Fine aggregate	30%

South of England, for both reinforced and prestressed concrete. It is important that shell and chloride content be limited. There are no recognized limits for shell content as the use to which the concrete will be put largely determines the acceptable limits. For high-grade, off-the-form concrete the presence of shell in the coarse aggregate may be prohibited, but in fine aggregate shell is comparatively unimportant, except that an excessive amount may increase the water demand of the mix. Table 5.2 gives the recommended quantities.

The chloride content of the aggregate is important because this will have a direct influence on the total chloride content of the concrete. The latter (the total chloride content of the concrete) is now strictly limited by the latest revision to Clause 6.3.8 of CP110. Table 5.3 contains the latest proposals of the British Standards Institution.

Table 5.3

Type or use of concrete	*Maximum total chloride content expressed as percentage of chloride ion by weight of cement*
Prestressed concrete, and Structural concrete that is steam cured	0·10
Concrete containing embedded metal made with sulphate resisting Portland cement (BS 4027), or super-sulphated cement (BS 4248)	0·20
Reinforced concrete made with ordinary or rapid hardening Portland cement	0·35 for 95% of test results with no result greater than 0·50

It can be seen that this revision to CP 110 virtually prohibits the use of accelerating admixtures based on calcium chloride except in plain concrete which does not contain embedded metal.

To convert sodium chloride content to chloride content, multiply by 0·61. The chloride salts in sea water are almost entirely sodium chloride; the conversion factor is based on atomic weights of Na = 23 and chlorine = 35·5. For conversion of calcium chloride, the factor is 0·64, based on the atomic weight of calcium of 40·0.

Limestone Aggregates

Limestone is a calcareous rock and consists essentially of calcium carbonate with which there is usually some magnesium carbonate and siliceous material such as quartz.

In dolomitic limestones the percentage of $MgCO_3$ is much higher and the rock contains the double carbonate $MgCO_3 . CaCO_3$.

There is a wide divergence of view on the suitability of limestone as an aggregate for concrete in water retaining structures, particularly when the water is from an upland gathering ground and is soft and slightly acid. Some engineers forbid the use of limestone for such structures, but it is seldom that this decision is based on sound technical grounds. The effect of soft acid waters on Portland cement concrete is discussed in detail in Chapter 6.

One advantage with limestone is that it has a lower coefficient of expansion than flint gravels and igneous rocks.

Artificial Lightweight Aggregates

As far as the authors are aware, structural lightweight aggregate concrete has not been used for what may be termed 'normal' liquid retaining structures in the UK. However, it can be used with advantage as a thermal insulating lining for tanks holding hot or very cold liquids and for the thermal insulation of roofs.

For structural concrete, the aggregates in use in the UK are sintered and expanded clays and shales and pulverized fuel ash, foamed blastfurnace slag, air-cooled blastfurnace slag, and clinker.

The bulk density and strength of concrete made from these aggregates vary according to the type of aggregate used, whether a natural sand is used as the fine aggregate instead of the fine lightweight material, and the mix proportions. Non-structural lightweight aggregates include pumice, vermiculite and perlite.

The special characteristics of all these lightweight aggregates are their thermal insulating properties, low density and, in some types, their sound absorbing properties.

5.4. *Steel Reinforcement*

General

Steel reinforcement for concrete including prestressed concrete is covered by seven British Standards which are listed in Appendix 1.

For normal reinforced concrete there are three types of reinforcing steel:

(a) Mild steel, plain and deformed and high yield deformed hot rolled bars (BS 4449).
(b) Cold worked steel bars (BS 4461).
(c) Steel fabric (BS 4483).

High tensile steel is identified in various ways; there are cold worked twisted bars where the twisting is clearly seen, cold worked ribbed bars, and hot rolled deformed bars. High tensile reinforcement without surface deformations is unlikely to be suitable for use at stresses higher than those approximate for mild steel due to bond and crack control requirements.

Galvanized Reinforcement

At the time of writing this book there is no British Standard for galvanized steel reinforcement, nor for galvanized prestressing wire, although both are used to a limited extent in the UK. It appears that in the US considerably greater use is made of galvanized reinforcement on the basis that the galvanizing reduces the risk of corrosion of the steel during the lifetime of the structure. It is a fact that unprotected steel is much more vulnerable to corrosion than Portland cement concrete. On the other hand, in a properly designed and constructed structure, the concrete should provide adequate protection to the steel for an indefinite period.

However, conditions sometimes arise when the use of galvanized reinforcement and prestressing wire is justified, and this is discussed later in this book.

Stainless Steel

There are three basic groups of stainless steel, namely

(a) martensitic,
(b) ferritic,
(c) austenitic.

Of these it is the austenitic steels which are most widely used in building and civil engineering, and the information which follows relates to this group.

Austenitic steel is an alloy of iron, chromium and nickel, and two types in this group contain a small percentage of molybdenum. The type most generally suitable for special use in water retaining structures is En58J (also known as 316 steel). This contains 18% chromium, 10% nickel and 3% molybdenum. It is very resistant to corrosion but is very expensive compared with mild steel. Although it can be welded easily it may be non-magnetic or only slightly so, and therefore it may not be possible to detect it with the normal type of electromagnetic cover meter. A further point is that mild steel in direct contact with stainless steel may suffer accelerated corrosion as it is anodic to stainless steel.

Stainless steel of the 316 type is covered by BS 1449: Part 4 for flat rolled strip, while bars are covered by BS 970: Part 4. The use of stainless steel is only justified in special cases where normal methods of protection of mild steel cannot be relied upon either due to difficulty in application or in the maintenance of the protective treatment.

Reference should be made to Chapter 6 for information on the use of stainless steel in concrete liquid retaining structures.

Non-ferrous Metals

Very few non-ferrous metals are used in the construction of concrete liquid retaining structures. Those that are used are mainly copper, phosphor-bronze and gunmetal. Some brief information of these metals and their use in liquid retaining structures is given in Chapter 6.

Aluminium is attacked by the caustic alkalis in damp concrete and therefore requires adequate protective coatings; it is seldom used in the type of structure covered by this book.

5.5. *Admixtures for Concrete*

General Considerations

A simple definition of an admixture is that it is a chemical compound which is added to concrete, mortar or grout at the time of mixing for the purpose of imparting some additional and desirable characteristic to the mix.

Admixtures are sometimes referred to as additives, but it is better to use the latter word for the addition of chemical compounds to cement at the cement works, where they are ground in at the time of manufacture.

Admixtures should only be used when they are really required to produce a particular result which cannot be obtained by normal mix design. They should not be used with any cement except ordinary and rapid hardening Portland cement without the approval of the cement manufacturer.

The following are the main purposes for which admixtures are used:

(a) To accelerate the setting of the cement and the hardening of the concrete, mortar or grout; these compounds are known as accelerators.
(b) To increase the workability of the mix (plasticizers). Also usually included in this category are water-reducing admixtures and 'waterproofers'.
(c) To retard the setting of the cement and slow down the rate of hardening of the mix. These are known as retarders.
(d) To entrain air in the mix (air entraining agents). These compounds give an air entrained mix, which should not be confused with an aerated mix. The latter is obtained by quite different compounds and is used for different purposes.
(e) Cement-replacement admixtures such as ground blastfurnace slag and pozzolanas such as pulverized fuel ash.

Accelerators

In the UK, until the early part of 1977, the active ingredient in commercial accelerating admixtures for concrete and mortar was calcium chloride. The position changed completely with the publication in May 1977 of Amendment No. 3 to CP110. The details of the latest proposals as they affect the use of calcium chloride are given in this chapter, Section 5.3, under 'Sea-dredged Aggregates'.

The main reason for the virtual prohibition of the use of chloride-containing admixtures is that the chloride ions are very aggressive to ferrous metals and can cause serious corrosion of reinforcement in concrete. The chloride ions do not attack Portland cement concrete itself, but they do reduce the sulphate resistance of sulphate resisting Portland cement. The serious effect of an excessive concentration of chlorides in reinforced concrete has been known for many years, and discussions on the subject have always revolved around the question of the 'safe' concentration which in fact is still largely unresolved as many factors are involved.

Because of the danger of using calcium chloride, considerable efforts have been made to find a satisfactory substitute as the basis for accelerators. It appears that so far, only two compounds have met with practical success, and these are calcium formate and sodium carbonate (washing soda). For many purposes, about 1% of anhydrous sodium carbonate by weight of cement will be adequate. However, in concrete liquid retaining structures, the use of accelerating admixtures is generally not recommended.

Retarders

There are two main uses for retarders. One is as an integral part of a concrete mix when it is required to extend the setting time of the cement and reduce the rate of hardening of the concrete. The other is when the retarder is used on formwork to retard the setting and hardening of the surface only of the concrete to facilitate work on the concrete surface when the formwork is removed.

Retarders are usually sugars and similar compounds, but borax is also used. The reaction between retarders and Portland cement is a very complicated one. It is affected by the chemical composition of the cement and the temperature of the concrete as it is maturing. Therefore the period of retardation can only be estimated approximately and accurate reproduction of results is usually impractical.

Air Entraining Admixtures

Air entrained concrete is seldom used for liquid retaining structures in the UK. Its main use is for external paving to resist the disintegrating effects of frost and de-icing salts. This type of admixture can be very useful in helping to provide a workable, cohesive mix.

The best air entraining agents are resins. They must be carefully dispensed into the mixing water, so that the dosage is accurately controlled and the compound is uniformly distributed throughout each batch. Their effect is to produce a large quantity of minute bubbles of air which alter the pore structure of the concrete. The effect of this entrainment of air (about $4\frac{1}{2} \pm 1\frac{1}{2}\%$) is to slightly reduce the compressive strength of the concrete but at the same time to improve the workability.

Air entraining agents should not be confused with such compounds as aluminium powder which is used for the production of aerated lightweight concrete and mortar.

Plasticizers or Workability Aids

These admixtures can be divided into two main types:

(a) Lignosulphonates, also known as lignins, and soaps or stearates.
(b) Finely divided powders.

The lignosulphonates and stearates act very largely as lubricants and in this way the amount of water required in the mix to obtain a predetermined workability can be reduced; or for a given water/cement ratio the workability is increased. Some of these compounds, mainly the stearates, impart a degree of water repellency to the concrete, mortar or grout and these are sometimes referred to as waterproofers. An overdose of these compounds will produce a set retarding effect and can result in a permanent reduction in compressive strength.

The finely divided powders include pulverized fuel ash, powdered hydrated lime, powdered limestone and bentonite. Portland cement itself is a good plasticizer and an increase in cement content may help to overcome problems of segregation and harshness. Depending on the characteristics of the powder and quantity used, the water demand of the mix may be increased but its cohesiveness may be improved.

To sum up, both types (the lignins and stearates, and powders) have their specific uses and both can be considered as reliable provided they are correctly used. They help to achieve a workable and cohesive mix which can be compacted under the action of poker vibrators and vibrating beams.

It should be remembered that many manufacturers of admixtures produce compounds which serve more than one purpose; this means that one can obtain plasticizers which also act as retarders, while other types act as accelerators. It is therefore important to obtain complete information on the basic composition and all the effects of a particular admixture before deciding to use it.

Superplasticizing Admixtures

These materials are a relatively new type of chemical admixture as far as their use in the UK is concerned. They have, however, been in commercial use in Japan since about 1967 and in Germany since about 1972. It is important to appreciate that superplasticizers are chemically distinct from the normal workability aids at present in general use; they can be used with confidence at high dosage levels subject to the general conditions set out in this section.

Superplasticizers can be used for two purposes:

(a) To produce a concrete having a virtually collapse slump, i.e. a 'flowable' concrete.
(b) To produce a concrete with normal workability but with a very low water/cement ratio, resulting in a high strength concrete.

It is not the intention in this section to discuss the chemical composition of the various types of superplasticizers available in the UK, but this can be summarized as follows:

Group 1—sulphonated Melamine formaldehyde;
Group 2—naphthalene sulphonated formaldehyde;
Group 3—modified ligno-sulphonates.

Of these, the first two groups appear to be the more reliable and effective.

Research work and site tests in Germany, Japan and this country, have shown that so far the use of these compounds has not revealed any adverse effects on the durability of the concrete, nor on the ability of the concrete to passivate steel reinforcement and protect it from corrosion, nor has reduction in the strength of the concrete over a long period of time been detected.

In order to obtain concrete which will 'flow' but not segregate, it is necessary to start with a slump of about 50–75 mm before the superplasticizer is added. The fine aggregate (sand) content of the mix must be increased by about 4–5 %, and the coarse aggregate correspondingly reduced so that the aggregate/cement ratio remains unchanged. The superplasticizer is added to the mix after the addition of the water and then mixing should be continued for at least another 2 min.

Correctly made superplasticized concrete has a slump of about 200 mm (when it is required to be flowable), and is almost self-levelling; it is cohesive and will not segregate.

Strict site control of the mix proportions, especially the sand content and the original slump, are essential if segregation is to be avoided.

Because of the danger of segregation when using these admixtures, it is most important that trial mixes be carried out prior to their use and to ensure that the finally decided mix design, including the type and grading of the fine and coarse aggregates, are exactly followed when the concrete is made.

It should be noted that maximum workability is usually only retained for a period of about 30–60 min, and then the concrete rapidly reverts to normal conditions of slump.

The information so far available on superplasticizers shows that they are a very useful addition to the concrete industry. One of the basic principles of good quality concrete in the structure is that it must be thoroughly compacted. Full compaction can often prove difficult and in some cases almost impossible to achieve if high strength is to be maintained. This is particularly so when repairs with concrete have to be carried out to walls and columns as well as members of small sections.

For further information on superplasticizers the reader is referred to the bibliography at the end of this chapter.

Pulverized Fuel Ash (PFA)

Pulverized fuel ash is produced in very large quantities as the residue from coal-burning, electricity-generating stations. It is a fine powder, having a similar specific surface to ordinary Portland cement, i.e. about 3500 cm^2/g. The specific gravity, however, is appreciably lower than that of cement, being in the range of 1·9 to 2·3 (ordinary Portland cement is about 3·12).

The main chemical compounds in PFA are oxides of silicon, iron and aluminium together with some carbon and sulphur. The relevant British Standard is BS 3892—*Pulverized Fuel Ash for Use in Concrete*. The Standard limits the amount of combustible material and sulphur compounds. Pulverized fuel ash possesses some pozzolanic properties and this is the principle reason why it is used in certain classes of concrete and grout.

A proposal to use PFA in structural concrete should take the following factors into consideration:

(a) Pulverized fuel ash should only be incorporated into concrete, grout or mortar with the written approval of the client's technical representative.
(b) For all structural concrete, a minimum cement content should be specified for reasons of durability, and the replacement of cement by PFA should not reduce the cement content to below this minimum figure.
(c) The chemical composition and the pozzolanic properties of PFA are closely related to the coal from which it is derived and the combustion characteristics of the power station where the coal is burnt. These variations can affect the compressive strength of the concrete, both at early ages and at later periods such as 3, 6 and 12 months, and therefore make quality control of the concrete difficult.
(d) The addition of PFA will affect the workability of the concrete, and it appears that the carbon content of the PFA is mainly responsible for this. The usual tolerance permitted in the specified slump for normal concrete is ± 25 mm, but if the concrete contains PFA it is likely to be found extremely difficult to maintain this tolerance. In addition to wider variation in slump, more rapid stiffening of the concrete may occur.

Consideration of item (b) above will show that this is more important than may appear at first sight. Minimum cement contents are based on experience over many years and these have a direct influence on durability. Durability is a very wide term and its interpretation depends on the requirements for the structure or part of the structure in question. The subject is discussed in detail in Chapter 6 where consideration is given to the durability aspects of the principal materials used in concrete liquid retaining structures.

When considering the use of admixtures of any type, including PFA as a substitute for part of the cement, the possible effect on the protection of reinforcement by the cement paste must be taken into account. The intense alkalinity of the cement (it has a pH about 12·0) provides protection (passivation) of the steel embedded in it. The pH of PFA is much closer to the neutral point of 7·0.

However, when PFA is used as a replacement for cement in massive concrete structures such as dams, it has proved to be very useful in reducing thermal effects during the early life of the concrete.

While the addition of PFA to a concrete mix made with ordinary Portland cement is likely to increase the sulphate resistance of the concrete, there are at present no officially published figures to show what the improvement is in specific sulphate concentrations. Therefore it cannot be considered as an effective substitute for sulphate resisting Portland cement.

Mention is made in Chapter 6 to recent cases of alkali–aggregate reaction in concrete in the UK. An essential factor in this reaction is the caustic alkali content of the cement, measured as equivalent Na_2O (sodium oxide). Recent work in Germany suggests that the total caustic alkali content of the concrete is possibly of equal importance. Therefore this should be taken into account when assessing the possible use of PFA. In the US there are official recommendations limiting the caustic alkali (equivalent Na_2O) content of PFA, but none at present in the UK.

5.6. *Joint Fillers and Sealants*

Fillers

Fillers for joints in liquid retaining structures are sometimes known as 'backup' materials. They are used in full movement joints with the object of providing a base for the joint sealant and also to prevent the ingress during the construction period of stones and similar which may prevent the joint from closing. Materials for these fillers include specially prepared fibres, cellular rubber, and granulated cork compounds. The material used should fulfil the following requirements:

(a) It must be very durable.
(b) It must be chemically inert, and when in contact with potable water it must be non-toxic.
(c) It must not afford a breeding place for fungi or micro-organisms.
(d) It must be resilient and should not extrude so as to interfere with the sealing compound, and should not bond to the sealant.
(e) It should be easily formed to the correct dimensions and be readily inserted into the joint.

Materials can now be obtained which fulfil the above conditions satisfactorily.

Joint Sealants

General Considerations

The materials used to seal joints in liquid retaining structures can be conveniently divided into two basic groups:

(a) Preformed materials.
(b) *In situ* compounds.

To be satisfactory, both groups must possess the following characteristics:

(a) The material itself must be impermeable to the liquid stored.
(b) As the joint opens and closes the sealant must deform in response to that movement without undergoing any change which will result in leakage. This integrity should be maintained throughout the working temperature range and throughout the whole life of the structure. This means that, dependent upon the type of joint in which it is placed, the sealant may be
    (i) permanently in tension, or
    (ii) permanently in compression, or
    (iii) alternately in tension and compression, or
    (iv) alternately or simultaneously in compression and shear (sliding joints).
(c) It must bond to the sides of the groove in which it is placed so that there is no leakage at the interface, but should not bond to the inert filler below.
(d) The material must be very durable as periodic renewal may be

difficult and expensive. Ideally, its life should be the same as the structure of which it forms a part.

(e) The material must be non-toxic in potable water tanks and should not form a breeding place for fungi and micro-organisms.

(f) It should be comparatively easy to apply under the weather and site conditions relevant to the location of the structure.

The sealant, whether preformed or *in situ*, is accommodated in a groove formed in the concrete. Work in both the UK and the US has shown that the shape and dimensions of the groove are important in ensuring a liquid-tight joint.

Preformed Materials

Preformed joint sealants at present take up only a small percentage of the market, but this share is likely to increase due to the superior durability of such materials as Neoprene and ethylene–propylene–diene–monomer (EPDM). The preformed material is mostly imported from the US and the Continent, and volume for volume is cheaper than high-quality *in situ* sealants such as polysulphides and silicone rubbers. However, the cost of accurately forming the groove to receive the preformed gasket reduces this margin.

When correctly installed, cellular Neoprene and EPDM gaskets will remain watertight against pressures up to 15-m head of water. The two materials mentioned are particularly resistant to sunlight and bacterial attack.

At the present time there is no British Standard for preformed sealants.

In Situ Compounds

In situ compounds can be divided into three main groups:

(a) Mastics.
(b) Thermoplastics (hot applied and cold applied).
(c) Thermosetting compounds (chemically curing and solvent release).

The authors are indebted to the American Concrete Institute for much of the information which follows.

Mastics

Mastics are generally composed of a viscous liquid with the addition of fillers or fibres. They maintain their shape and stiffness by the formation of a skin on the surface and do not harden throughout; the material does not set in the accepted use of the term. The vehicles (i.e. the viscous liquids) are usually low melting point asphalts or polybutylene, or a combination of these. They are used where the overriding factor is low first cost, and maintenance and replacement costs are not considered important. The extension–compression range is small and so these materials should only be used where small movement is anticipated.

Thermoplastics—Hot Applied

These materials become fluid on heating, and on cooling they become an elastic solid, but the changes are physical only and no chemical reaction occurs.

A typical example of this type of sealant is the rubber-bitumen compounds which are used extensively in many countries.

As the sealant has to be applied in a semi-liquid state it is only suitable for horizontal joints; it is used largely for roads and airfield pavements, but can also be used in the floors of reservoirs and sewage tanks.

The movement range which this type of material can accommodate is rather greater than that obtained with mastics, but is still small compared with thermosetting chemically curing elastomers.

There is a British Standard, BS 2499—*Hot Applied Sealing Compounds for Concrete Pavements*. The Standard is strictly a performance specification and says nothing about the chemical composition of the sealants. When used for tanks to hold potable water the sealant must be non-toxic and should not contain phenol compounds.

Thermoplastics—Cold Applied

These materials set and harden either by the evaporation of solvents (solvent release), or by the breakup of emulsions on exposure to air. Sometimes a certain amount of heat is applied to assist workability, but generally they are used at ambient air temperature.

For water retaining structures the most popular type is a rubber-asphalt. The movement which this type of sealant can accommodate is small; it hardens with age and suffers a corresponding reduction in elasticity. There is no British Standard for this type of sealant. A check should be made to ensure that the sealant does not contain toxic or phenol compounds before accepting its use in potable water tanks.

Thermosetting Compounds—Chemically Curing

Materials in this category are one- or two-component compounds which cure (mature or harden) by chemical reaction to a solid state from the liquid or semi-liquid state in which they are applied.

High-grade materials in this class are flexible and resilient and possess good weathering properties; they are also inert to a wide range of chemicals.

These compounds include polysulphides, polyurethanes, silicone rubber and epoxide-based materials. They can be obtained so as to have an expansion–compression range of up to ±25% and a temperature range from −40°C to +80°C.

While they are considerably more expensive than the mastics and thermoplastics, they will accommodate far greater movement and are more durable.

In the UK the polysulphides are the most popular and are used in a wide range of liquid retaining structures, including those containing potable water. Some are used with a primer and some without. It is important to ascertain whether the particular brand selected will bond to damp concrete or whether a dry surface is required. Complete adhesion between the sealant and the sides (but not the base) of the sealing groove is essential for a liquid-tight joint. In the climatic conditions of the UK it is virtually impossible to ensure dry concrete on the general run of construction sites. Two-part polysulphide sealants are covered by BS 4254, which deals with two grades, a pouring grade and a gun grade. There is no other British Standard for any of the other types of thermosetting sealants.

Thermosetting Compounds—Solvent Release

Sealants of this type cure by the release of solvents present in the compound itself. The principal materials in use are based on such compounds as butyl, Neoprene and polyethylene.

Their general characteristics are somewhat similar to those of solvent-release thermoplastics; their extension–compression range is about ±7%.

There is no British Standard for this type of sealant.

5.7. *Water Bars*

There are two categories of water bars in general use, namely rigid and flexible.

The two materials used for rigid water bars are copper and steel but neither are used much these days. There is no British Standard for metallic (rigid) water bars for use in liquid retaining structures. The nearest equivalent is BS 1878—*Corrugated Copper Jointing Strip for Expansion Joints for Use in General Building Construction*. Steel water bars can be mild steel, plain or galvanized, or stainless steel. Apart from dams, which are not dealt with in this book, the only locations in which a rigid water bar should be used is between the kicker and the wall panel, and in horizontal joints between 'lifts' in high walls. While, as stated above, there is no British Standard for copper water bars for liquid retaining structures, there may be occasions when the use of copper is considered desirable. The material has been used in the past in many parts of the world and has proved to be resistant to corrosion, and is durable, provided it is not subjected to reversals of stress over a long period. Fatigue takes the form of hardening, and brittle fracture may occur. The thickness range in BS 1878 is 0·022–0·032 in (0·5–0·75 mm). Care must be taken to ensure that the copper is not in contact with any other dissimilar metal. Reference should be made to the British Standards Institution document PD 6484—*Commentary on Corrosion at Bi-metallic Contacts and its Alleviation.*

Non-metallic flexible water bars are by far the most widely used. The principal materials are natural rubber and polyvinyl chloride (PVC). Other materials which can be used for special purposes include butyl rubber, polyisobutylene, Neoprene, and styrene butadiene rubber. All types should possess the following characteristics:

(a) They should be very durable under the operating conditions.
(b) They should be non-toxic when in structures containing drinking water or other potable liquids.
(c) They should be inert and not form a breeding place for fungi and micro-organisms.
(d) They should be sufficiently rigid to maintain their position and shape during installation and during placing and compaction of the concrete.
(e) They should bond effectively to the concrete in which they are embedded.
(f) They should possess the necessary degree of flexibility and elasticity for the operating conditions of the joint which they have to seal. Usually movement at joints is largely cyclic in response to environmental changes and this may result in periodic reversal of stress.
(g) The joints should be made with solvent welding.

Proprietary water bars of both rubber and PVC fulfil the above conditions reasonably well with the exception of (e), good bonding with the surrounding concrete. The fixing of the water bar in the concrete is obtained by the mechanical lock of the hardened concrete around the outside of the bar. An exception to this is a Dutch water bar which contains a mild steel tongue at each end, while the bar itself is made of artificial rubber; it is comparatively expensive.

Rubber water stops possess superior tensile strength and elongation characteristics compared with PVC, but the latter can be used with confidence for a wide range of conditions; the appreciable extra cost of rubber is only justified in special cases. Typical figures for good quality rubber and PVC are given in Table 5.4.

Table 5.4

Material	*Tensile strength*	*Elongation*
Rubber	21 N/mm^2 (3 000 lbf/in^2)	450 %
PVC	14 N/mm^2 (2 000 lbf/in^2)	300 %

Both rubber and PVC water stops can be classified according to their shape and their position in the structure. Those which are located centrally within the cross-section of the floor, wall or roof are usually either plain dumb-bell shape or centre-bulb dumb-bell; a third shape which is seldom used in the UK is the flat corrugated type. Water bars which are located on the surface of the joint are known as surface or external types.

The actual use of the various shapes is discussed in Chapters 7 and 8.

5.8. *Organic Polymers*

General Considerations

Organic polymers are complex chemical compounds derived mainly from the petrochemical industry. These materials are often referred to as 'resins', and the principal resins used in the construction industry are epoxide, polyurethane, polyester, acrylic, polyvinyl acetate, and styrene butadiene.

The basic raw materials are supplied by comparatively few manufacturers, such as Shell Chemicals, CIBA, Dunlop Chemical Products Division, Revertex, Borden Chemical Co., and Dow Chemicals. The raw materials are then taken by a large number of formulators who formulate the final products in such a way that they possess specific characteristics needed for the use to which they will be put.

Under the particular condition of curing, which in some cases requires hardeners or accelerators, the resins form long molecular chains in three dimensions, which can result in an extremely strong and stable material.

While the range of use of these materials is now very wide, it is convenient and practical to divide it into two main categories; namely coatings in which the formulated compound is used on its own, and mortars and concretes in which the resin is mixed with aggregate and sometimes cement.

It is usual to find that more than one type of polymer is used in combination in order to obtain the optimum results. The information given here is intended for practising engineers and not for chemists.

For coatings used for the protection of concrete and to improve impermeability, epoxies and polyurethanes are in most general use. For use in mortars and concretes, epoxies, polyurethanes, and acrylic and styrene butadiene compounds are used successfully. Polyvinyl acetate is used in large quantities as a bonding agent for floor screeds and toppings to increase adhesion with the base concrete; it is also used in cement mortar mixes to improve certain characteristics of the mortar and bond with the substrate.

Epoxide Resins

The resins as marketed by the formulators have the special properties required for the specific use to which they will be put. For example, some resins can be successfully applied and cured under water. While most epoxies are rigid when cured, it is now possible to obtain a type which is slightly flexible.

The basic characteristics of epoxide resins include the following:

(a) Outstanding adhesive qualities to such materials as concrete and steel.
(b) Resistance to a wide range of acids and alkalis and other chemicals.
(c) Rather vulnerable to organic solvents.
(d) Low shrinkage when the compound cures and changes from the liquid to the solid state.
(e) High coefficient of thermal movement compared with concrete.
(f) High compressive, tensile and flexural strength.
(g) Appreciable loss of strength at temperatures over about 80 °C.
(h) Low viscosity if required.
(i) The reaction between the resin and the accelerator is exothermic.

Reference is sometimes made to 'water miscible' or 'water soluble' resins. Only a very limited range of epoxies can be successfully mixed with water. Such resins are generally used for making epoxy mortars. The advantage is cheapness; the disadvantage is relatively low strength compared with a normal resin. The water is usually mixed with the resin, the activator is then added, followed by the fine aggregate. Such a mortar has about the same strength as an SBR mortar, but due to the high cost of the epoxy resin, the overall cost of the mortar is higher. Chemical resistance may be a little better than an SBR mortar.

Solvent Release

Most epoxies and polyurethanes contain a solvent or a mixture of solvents. This greatly facilitates the application of the resin by spray or brush.

The type of solvent(s) is determined by the formulator and is influenced by the ambient application temperature and the curing time required.

The solvents evaporate after application and this is why the dry film thickness is appreciably less than the wet film thickness.

If the wrong solvent is used pin holes can form in the cured resin. The blending of solvents is very important and highly specialized.

Solvent Retention

Formulators can arrange that not all the solvent is released at once. Different solvents have different boiling points. For example, 90 % may be released fairly quickly, and the remainder only at a higher temperature.

Polyurethanes

This useful range of polymer resins can be obtained as one- or two-pack systems. In general, the two-pack system will give better durability than the one-pack system.

Correctly formulated two-pack polyurethanes when used with an appropriate primer give very good abrasion resistance, and like most polymer resins are resistant to a wide range of chemicals, but are liable to deterioration when in contact with acetic and formic acids, ammonia and acetone. The coating material can be formulated to give a high degree of flexibility and is therefore suitable for structures subject to thermal movement.

Some formulations of polyurethanes polymerize on contact with moisture contained in concrete. While this has advantages, the rate of polymerization is unpredictable, and not infrequently results in the formation of blisters.

An important characteristic of polyurethanes is that they can be formulated so as to be colour-fast which many other resins cannot.

There is now a decided tendency to use polyurethanes instead of epoxides and polyesters. Some experienced formulators produce resins for special purposes which consist of a polyurethane hardener (the catalyst), and an epoxide base resin.

Styrene Butadiene and Acrylic Resins

These materials are also known as latexes and polymer emulsions.

Some of the properties of a styrene butadiene copolymer emulsion (latex), used with Portland cement, in grout, mortar and concrete are

pH	11·0
Total solid content	47%
Specific gravity	1·1

The authors are indebted to Revertex Ltd for the above information which relates to their latex 'Revenex 29Y40'.

Acrylic latexes (styrene acrylic and all acrylic) generally have shorter setting and hardening times and greater resistance to ultra-violet light than the styrene butadiene type, but are more expensive.

The advantages claimed for the use of these latexes in Portland cement grout, mortar and concrete include reduced permeability, reduced initial drying shrinkage, improved resistance to attack by certain dilute acids and solutions of sulphates, and improved bonding to the substrate.

Polyester Resins

Polyester resins are used mainly with Portland cement and selected aggregate to form a polymer–cement–aggregate mortar. Such mortars possess a number of desirable qualities such as good resistance to a wide range of chemicals, high resistance to abrasion, complete watertightness under the range of heads likely to be met in liquid retaining structures, and high bond strength with most common building materials.

There are a number of important differences between the properties of polyester resins and epoxide resins. These include wider temperature range, better resistance to heat, and higher shrinkage, but bonding properties to concrete are generally lower. However, by adjusting the ratio of resin to catalyst the hardening time can be made very short, and once solidification occurs the rate of gain of strength is very rapid.

One of the main uses of this resin is in the formulation of proprietary floor toppings with special properties. One of the best known and most successful is 'Estercrete'.

Polyvinyl Acetate (PVA)

Polyvinyl acetate is used as a bonding agent and as an admixture in mortar for screeds to improve certain properties of the mortar. Manufacturers of proprietary compounds based on PVA claim increased tensile and flexural strength, reduced drying shrinkage, and reduced permeability. The authors consider that some of the improvements claimed are marginal and the behaviour of PVA under permanently damp conditions can be very disappointing.

Synthetic Rubbers

The first synthetic rubber to be widely used in industry and commerce was Neoprene which was introduced by du Pont in 1932. Since then a wide range of synthetic rubbers have come on to the market, each with its own special properties; the range includes such well-known materials as butyl rubber, polyisobutylene, 'Hypalon', EPDM, etc. Nitrile rubber and 'Viton' are among the latest materials in this constantly expanding range of synthetic rubbers. Many of these materials can be obtained in both liquid and sheet form. EPDM has improved characteristics over Neoprene such as greater elasticity and resistance to fatigue stress.

In water retaining structures, EPDM is used for preformed cellular gaskets for sealing joints. It exhibits outstanding resistance to sunlight, weathering, bacterial attack, a wide range of chemicals, heat and abrasion. 'Hypalon' can be obtained in liquid and sheet form, and in a wide range of colours. Butyl rubber and polyisobutylene are available only in sheets and in one colour, black. The sheets can be solvent-welded, and polyisobutylene can be effectively bonded to properly prepared concrete.

5.9. *Polymer Concrete*

In this context, polymer concrete means Portland cement concrete which is polymerized *in situ* and this should not be confused with concrete in which the cement is replaced by an organic polymer such as epoxide or polyester resin.

Polymerized concrete can be divided into two basic categories:

(a) In which the complete precast concrete unit is impregnated (usually by dipping) with a monomer and is then polymerized by either heat or gamma rays.

(b) In which a monomer is mixed into the gauging water used for the concrete and then after the concrete has hardened to a predetermined figure it is polymerized by heat.

Claims for polymerized concrete include an increase up to four times in the compressive and tensile strengths and a considerable increase in resistance to chemical attack and freeze–thaw conditions, while absorption and permeability are substantially reduced. However, these improvements are at present accompanied by a substantial increase in cost and there is little published information on the practical use in the field of this type of concrete.

Considerable research and development work on polymer concrete is going on in many countries and it is hoped that eventually production costs will be appreciably reduced, so that its special properties can be utilized for practical purposes.

5.10. *Fibre Reinforced Concrete*

The first significant research work on fibre reinforced concrete was apparently carried out by Romualdi and Batson and published in 1963. This work was principally concerned with the effect of random steel wire on the crack pattern in concrete. There are still differences of opinion on the reasons for the observed effects of the introduction of short random fibres into a concrete mix, and on the extent to which this improves the strength of the concrete.

As far as the authors are aware, fibre reinforced concrete has not been used for liquid retaining structures, but this does not mean that it will never be used for this purpose.

The fibres on which most development work has been concentrated are steel, glass and polypropylene. Of these, steel appears to give the greatest improvement, but the presence of the fibres close to the surface of the concrete gives rise to rust staining which in some cases is not acceptable. This staining has no ill effect on the durability of the concrete.

Ordinary glass is attacked by caustic alkalis and therefore suffers deterioration from the intense alkalinity of the cement paste. A special alkali-resistant glass has been developed, but some authorities feel that the regression in strength over a long period of time is still not subject to sufficiently accurate assessment to enable the composite material to be used for important structural purposes.

Polypropylene fibres have been found to impart appreciable increase in resistance to impact and have been used successfully in the concrete shells of a proprietary piling system.

Work has been carried out in the US and elsewhere on the use of glass fibres protected with epoxide resins of various compositions. Published information on this work shows that for normal (corrosion-susceptible) glass fibres used with Portland cement, the fibres must be completely and permanently protected from the alkaline environment of the cement paste. In addition to protection the coating must provide a good surface bond between the fibres and the surrounding mortar or concrete.

5.11. *Ferro-Cement*

Ferro-cement is a material used principally in boat building and was originated in the early 1940s by Piere Luigi Nervi. Professor Nervi's description of this new material is still valid and is given below:

> ...The fundamental idea behind the new reinforced concrete material, ferro-cement, is the well known and elementary fact that concrete can stand large strains in the neighbourhood of the reinforcement, and that the magnitude of the strains depends on the distribution and subdivision of the reinforcement throughout the mass of concrete. With this principle as a starting point, I asked myself what would be the behaviour of slabs in which the proportion and subdivision of the reinforcement were increased to a maximum by surrounding layers of fine steel mesh, one on top of the other, with cement mortar....

Ferro-cement consists of a cement/sand mortar with a high percentage of reinforcement. The mix proportions are about 1 part Portland cement and 1·5 or 2 parts clean sharp sand, with a water/cement ratio of about 0·35. The percentage of reinforcement is about 15% by weight. The mortar including the reinforcement has a density of about 2700 kg/m^3(165 lb/ft^3). The mortar is applied by hand to the reinforcement which consists of a relatively fine mesh, or layers of mesh, which is sometimes securely wired to vertical and horizontal main reinforcing bars. The amount and the detailing of the reinforcement depends largely on the thickness of the unit and on the judgement of the designer. Very often galvanized mesh is used.

The authors have been unable to find any published record of the use of ferro-cement for liquid-containing structures. However, in view of its undoubted success as a material for the hull of boats it should be satisfactory for liquid retaining structures of limited size. When properly executed, ferro-cement is resistant to the passage of water and to damage by impact. Although the cover to the steel is very small by the standards adopted for concrete, it is a known fact that corrosion of the reinforcement does not occur when the work is properly carried out.

A recent development is to use alkali-resistant glass fibre mesh instead of steel; another is to use polypropylene fibre.

The authors consider that in the next decade there will be considerable development in ferro-cement techniques and an appreciable extension in its use.

Bibliography

American Concrete Institute. Admixtures for concrete. *J. ACI*, **60**(11), November 1963, pp. 1481–1526.

American Concrete Institute. Guide to joint sealants for concrete structures. Title 67–31. *J. ACI*, July 1970, pp. 489–536.

British Standards Institution. *Commentary on Corrosion of Bi-metallic Contacts and its Alleviation*. PD 6484, London, 1979, 26 pp.

British Steel Corporation. *Building with Steel*, No. 9, February 1972. Theme—Anti-corrosion, pp. 2–22.

Brown, B. L., Harrop, D. and Treadaway, K. W. J. *Corrosion Testing of Steels in Reinforced Concrete*. Current Paper CP 45/78, Building Research Station, May 1978, 12 pp.

Building Research Establishment. *Shrinkage of Natural Aggregates in Concrete*. Digest No. 35 (Second Series), June 1963, HMSO, London, 5 pp.

Building Research Establishment. *Stainless Steel as a Building Material*. Digest No. 121, September 1970, HMSO, London, 4 pp.

Cahn, C., Phillips, J. C., Ishai, O. and Aroni, S. Durability of glassfibre–Portland cement composites. Title 70–18. *J. ACI*, March 1973, pp. 187–9.

Canadian Portland Cement Association. *Design and Control of Concrete Mixtures*. Portland Cement Association, Skokie, USA, 1978, 131 pp.

Cement and Concrete Association and Cement Admixtures Association. *Superplasticizing Admixtures for Concrete*. Cement and Concrete Association, London, 1976, 32 pp.

Collen, L. D. G. *Some Experiments in Design and Construction with Ferro-Cement*. Institution of Civil Engineers of Ireland, January 1960.

Concrete Society. *Admixtures for Concrete*. Technical Report TRCS1, London, December 1967, 10 pp.

Figg, J. W. and Lees, T. P. *Field Testing of the Chloride Content of Sea-Dredged Aggregates*. Reprint 3/75, Cement and Concrete Association, London, 3 pp.

Fletch, K. E. *Admixtures for Concrete*. Current Paper CP 3/74, Building Research Station, Garston, 5 pp.

George, C. M. *The Structural Use of High Alumina Cement Concrete*. Lafarge Fondu International, France, 1975, 15 pp.

Guera, A. J., Naaman, A. E. and Shah, S. P. Ferro-cement cylindrical tanks; cracking and leakage behaviour. *J. ACI*, January 1978, pp. 22–30.

Hoff, C. H. and Houston, B. J. Non-metallic water stops. Title 70–2. *J. ACI*, January 1973, pp. 7–13.

Lea, F. M. *The Chemistry of Cement and Concrete*, Third Edition. Edward Arnold Ltd, London, 1970, 725 pp.

Lowrey, K. W. *Protective Coatings for Structural Systems*. Civil Engineering UK Reprint, October 1972, 7 pp.

Midgley, H. G. and Midgley, A. The conversion of high alumina cement. *Magazine of Concrete Research*, **27**(91), June 1975, pp. 59–77.

Monks, W. L. *The Performance of Water Stops in Movement Joints*. Code 42.475, Cement and Concrete Association, London, October 1972, 8 pp.

Palmer, D. *Alkali–Aggregate (Silica) Reaction in Concrete*. Code 45.033, Cement and Concrete Association, London, Advisory Notes, December 1977, 8 pp.

Pomeroy, C. D. *The Influence of the Aggregate on the Mechanical and Other Properties of Concretes of Specified Cube Crushing Strengths*. Code 42.540, Cement and Concrete Association, London, January 1975, 16 pp.

Pomeroy, C. D. Concrete, an alternative material. *Fourteenth John Player Lecture, Institution of Mechanical Engineers, Proceedings*, **192**(14), 1978, pp. 135–44.

Robson, T. D. *High Alumina Cements and Concretes*. Contractors Record Ltd, London, and John Wiley & Sons Ltd, New York, 1962, 263 pp.

Romualdi, J. P. and Batson, G. B. Behaviour of reinforced concrete beams with closely spaced reinforcement. *J. ACI*, **60**(6), June 1963, pp. 775–90.

Romualdi, J. P. and Batson, G. B. Mechanics of crack arrest in concrete with closely spaced reinforcement. *Proc. ASCE*, **89**, EM3, June 1963, pp. 147–68.

Shirley, D. E. Principles and practice in the use of high alumina cements. *Municipal Engineering*, **145**(4 & 5), 1968.

Tabor, L. J. *Effective Use of Epoxy and Polyester Resins in Civil Engineering Structures*. CIRIA Report No. 69, January 1978, 70 pp.

CHAPTER 6

FACTORS CONTROLLING THE DURABILITY OF CONCRETE

6.1. *Introduction*

The intention of this chapter is to consider the more important problems which arise during the design and specification stage of a project, except the actual structural calculations which are dealt with in Part 1 of this book.

It is not possible to place design, specification and construction in completely separate compartments, and in the opinion of the authors this would, in any case, be a most unsatisfactory procedure.

It is essential that the design process (pre-construction stage) should be closely related to construction techniques and procedure on site. A simple example is the question of joints in a liquid retaining structure. While this is an essential part of the design process, it is discussed in detail in Chapter 7 because it is so closely related with construction on site.

It is important that the engineer responsible for the preparation of the contract documents, drawings, specification and general conditions of contract should have had considerable site experience and have a sound knowledge of the characteristics of the materials which will be used in the structure.

6.2. *Protective Measures for Structural Materials (Metals and Concrete) in Aggressive Environments*

General Considerations

At the present time there is no material known which is completely inert to chemical action and immune to physical deterioration. While concrete is no exception to this, it has, under what may be termed 'normal conditions of exposure', a very long life. Concrete made from naturally occurring cements (pozzolanic materials such as trass) has been found in excellent condition after more than 2000 years. There is no reason to believe that under similar conditions, modern Portland cement concrete will have a shorter life. This should satisfy even the most conservative client! However, it is known that environmental conditions can be aggressive to concrete and other materials used in modern construction.

Only two basic materials will be considered in this section, namely metals and concrete used in liquid retaining structures. The metals most commonly used are various types of steel (ferrous metals), copper, phosphor-bronze, gunmetal, zinc and aluminium.

For the purpose of this section, the term 'aggressive environment' is intended to mean any environmental conditions under which the structure will operate which may cause a deterioration in the ability of the structure to fulfil its original purpose and as a consequence may reduce its useful life. It does not include errors in design, overloading, and foundation failure, and normal weathering.

Metals

Mild and High Tensile Steel

The metal most used in liquid retaining structures is mild steel or high tensile steel, as reinforcement.

Reinforcing steel is protected by the cement matrix in the surrounding concrete, and emphasis must be placed on obtaining a dense impermeable concrete cover to the steel.

Steel embedded in Portland cement concrete is in an intensely alkaline environment which is created by the hydration of the cement. The pH of the cement matrix is in the range 11·5–12·5. Research has shown that within this pH range iron is passivated and as long as this is maintained, corrosion is inhibited. The corrosion of steel is a complicated electrochemical process. Unprotected steel corrodes in the presence of oxygen and moisture due to differences in electrical potential on the surface of the steel which forms anodic and cathodic areas. The metal oxidizes at the anode and corrosion occurs there. Treadaway and Russell in BRS Current Paper CP 82/68 give the following three factors as determining the corrosion rate of steel reinforcement in concrete:

(a) Contact between the steel and the ionically conducting aqueous phase in concrete; dependent upon moisture content and constituents of the concrete.

(b) The presence of anodic and cathodic sites on the metal in contact with the electrolyte. This is a function of metal surface variations (oxide coating in contact with bare metal surface) and the environment.

(c) The availability of oxygen to enable the cathodic reactions to proceed (cathodic depolarization).

Cathodic depolarization is controlled by the rate of diffusion of oxygen from the atmosphere through the concrete matrix to the cathodic sites on the steel surface. From this it can be seen that the extent of corrosion, or rate of corrosion, is partly dependent on the permeability of the concrete. This applies equally to the reduction of the pH of the concrete by the reaction of carbon dioxide with the alkalis in the cement paste. If the concrete carbonates to a depth equal to the concrete cover to the reinforcement the passivity of the steel will be destroyed and corrosion accelerated. It has been estimated that if the pH falls to 9·0 corrosion is likely to start.

Passivity can also be destroyed by the presence of chloride ions and thus the presence of chlorides in a concrete mix is a potential cause of corrosion. The corrosion potential of chlorides is increased by porous concrete, which in turn can result from inadequate compaction, high water/cement ratio, low cement content, or poor mix design. The natural passivity of the concrete surrounding the reinforcement can be increased, and therefore its protection improved, by the use of corrosion inhibitors.

Corrosion inhibitors can be incorporated in the concrete, but in fact this course is not recommended for a number of reasons which are discussed in the Current Paper mentioned above. A better method is to include the selected inhibitors in a cement grout which is applied as a coating to the reinforcement. The work done by Treadaway and Russell suggests that a combination of sodium nitrite and sodium benzoate is likely to give the best results using readily available and fairly inexpensive chemicals. The authors' experience is that cement grout applied to reinforcement tends to dry rather quickly and flake off, but this can be substantially improved by the substitution of a styrene butadiene latex for the gauging water. Even so, it is important that the concrete should be placed as soon as practical after the coating of grout has hardened.

An alternative, more durable, but more expensive coating is epoxide resin. However, it would only be in extreme cases that such materials would be justified. This is discussed in some detail later in this chapter.

The authors consider that a short discussion is justified on the subject of the removal or otherwise of mill-scale and/or rust, on reinforcement.

Mill-scale is formed under mildly reducing conditions during the manufacture of the bars and consists of oxides of iron, with the lower oxides, FeO and Fe_2O_3, predominating. The higher oxide, Fe_3O_4, is also likely to be present. Mill-scale is defined in BS 5493—*Protection of Iron and Steel Structures from Corrosion* as 'stratified layers of the oxides of iron—FeO, Fe_2O_3 and Fe_3O_4'. As the mill-scale weathers it combines with oxygen in the air, i.e. it oxidizes and increases in volume and tends to lose its adhesion to the steel below. Generally, mill-scale is not continuous over the whole bar but is present as patches. It is this discontinuity which assists the formation of corrosion cells due to the difference in electrical potential between the mill-scale and the steel.

The question now arises as to what extent it is necessary to remove mill-scale from reinforcement. This is a subject which causes fierce arguments on site and is one on which it is not possible to lay down hard and fast rules, as the best guide is common sense. Having said this, the authors recommend that all *loose* mill-scale should be removed; in fact a great deal of it will probably have come off during handling and transportation. There is no need to remove mill-scale which is adhering firmly to the bars; the common sense mentioned above is needed to decide what is loose and what can be safely left.

Similar principles can be applied to rust on reinforcement. Light, powdered rust which does not come off when the bars are hit can be left; in fact some engineers claim that this type of rust is likely to improve bond to the hardened concrete; for the reasons given above, the presence of rust will slightly increase the risk of corrosion if the passivity of the cement matrix is reduced. To what extent, if any, the presence of this type of rust will improve or reduce bond during the early life of the concrete is open to question. While there are a number of reliable reports on the effect of rust on bond to hardened (mature) concrete, the authors have not seen any published work on bond to rusty steel during the early life of the concrete, i.e. during the first 12–72 h after casting.

Rust in the form of scales or flakes should be removed before concrete is placed.

Methods of protection for exposed steelwork are generally well known, but what is often not fully appreciated is the need for the most meticulous preparation of the steel surface prior to the application of the protective coating. These coatings can be in the form of galvanizing, metal spray (usually zinc), epoxide and polyurethane resins, plastics, or a paint system.

The standard of finish of the steel prior to the application of the protective coating is 'first quality (white metal) finish' as described in

BS 5493, and reference can usefully be made to this Code for a great deal of essential information and advice on this subject.

Steel which has been protected by galvanizing or metal spray is not usually painted in addition, unless the environmental conditions are very severe or maintenance is particularly difficult.

Galvanized Reinforcement

It must be remembered that moist Portland cement concrete, because of its alkalinity, will attack zinc and aluminium. It is therefore advisable to protect with bituminous paint or similar coating any aluminium or zinc coated steel which is in continuous contact with damp concrete.

When galvanized reinforcement is used in concrete, there will be some initial attack on the galvanizing, but this is unlikely to be significant in good quality, impermeable concrete. Some Portland cements contain chromates; if the chromate content exceeds about 65 ppm, this will inhibit the reaction between the cement paste and the zinc by the formation of a passive layer of zinc chromate on the surface of the galvanizing. If chromates are not present in the cement, then it is relatively simple to add them in the galvanizing vats.

Stainless Steel

Stainless steel has been discussed briefly in Chapter 5 and all that will be said here is that it is important to select the right grade for the anticipated conditions of exposure. For severe conditions it is advisable to specify type En58J (also known as type 316 steel), which is an austenitic steel; the normal use is for dowel bars and fixings.

It is prudent to take expert advice, from such an organization as the Corrosion Advice Bureau of the British Steel Corp., when it is necessary to have two metals of different chemical composition in contact, as this can give rise to electrolytic action resulting in the corrosion of the metal which forms the anode. Magnesium, zinc and aluminium are anodic to iron and steel; but iron and steel are anodic to brass, copper, phosphor-bronze and gunmetal.

Aluminium

If unanodized aluminium is used in continuous or frequent contact with damp concrete it should be protected by a thick coating of bituminous paint as it is attacked by the caustic alkalis in the cement.

The durability, appearance and resistance to corrosion of aluminium can be improved by anodizing. This is a process in which aluminium is coated with layers of oxide by making it anodic in a special solution. An alternating electric current is passed through the solution and a fairly thick coating can be built up. The process is carried out in a bath containing either chromic acid or sulphuric acid; the latter is used when it is required to produce a pigmented coating.

The protective value of the coating lies in the fact that aluminium oxide is inert and relatively impermeable. The impermeability and hence durability of the coating can be improved by sealing, which consists of immersion in hot water containing chromates. This hydrates part of the aluminium oxide and seals the surface pores. British Standard BS 1615—*Anodic Oxidation Coatings on Aluminium* gives methods of testing this type of coating.

Copper

Copper is resistant to attack in most conditions met in water and sewage works. It is not corroded by Portland cement concrete unless chlorides or ammonium compounds are present.

If copper is used as a water bar, or for pipes passing through the structure, care must be exercised to ensure that chloride-free concrete is used and that the copper does not come into contact with steel reinforcement or other metals used for fixings. Sea-dredged aggregates should be checked for chloride content; see the latest revision to CP110.

Phosphor-Bronze and Gunmetal

These two alloys are often used for fixings in structures containing aggressive liquids or where regular maintenance is not practical.

Bronze is a copper–tin alloy, and phosphor-bronze usually contains phosphorus as a copper phosphide.

Gunmetal is bronze which contains about 9% tin.

Both phosphor-bronze and gunmetal are inert to Portland cement.

Concrete

Durability and Impermeability

General Considerations

In concrete, durability and impermeability are not necessarily connected. An example is a no-fines concrete tennis court or play area. On the other hand an impermeable concrete is likely to be very durable unless it is attacked by an aggressive environment or suffers some form of physical damage.

However, it is a usual requirement that concrete liquid retaining structures should be both durable and possess a high degree of impermeability.

The authors have often met divergent views on the practical meaning of these two words and as these are key words in any book on liquid retaining structures, a short discussion on this subject follows.

Durability

A structure should be considered durable if it fulfils its intended duty for the whole of its design life with the minimum of maintenance. The design life of the structure will usually be decided by the client in consultation with the designer.

It would be unrealistic to expect any structure to maintain its 'as new' condition for a period of 60–100 years without any maintenance whatever. However, Portland cement concrete has the potential of an almost unlimited life unless it is subjected to chemical attack by an aggressive environment, or suffers physical damage. Weather staining and similar discoloration should not be confused with lack of durability. On the other hand, cracking and spalling of the concrete, due to poor quality materials or workmanship, and/or corrosion of the reinforcement, would be a clear case of low durability.

Concrete liquid retaining structures are not 100% concrete; there is reinforcement, joint sealants and fillers, metal fixings, pipe connections, and often waterproofing layers on the roof. Some of these other materials will have a limited life compared with the concrete. In particular, joint sealants and waterproofing layers require periodic renewal.

In some cases, certain parts of a liquid retaining structure may be subject to physical aggression, such as abrasion, when a jet or stream of high velocity water, possibly containing grit, impinges on a concrete wall or floor. In such a case, special measures would have to be adopted to ensure maximum practical durability, and this is considered later in this chapter.

Concrete in very exposed positions in the north of the UK and on high ground may be subjected to frequent and prolonged freeze–thaw cycles. This can cause surface spalling unless the concrete mix has been air entrained.

Impermeability

The standard of impermeability required for a concrete liquid retaining structure will be governed by the following:

(a) The type of liquid stored. A higher standard is likely to be required for a tank holding an inflammable or corrosive liquid than for one holding water or sewage. Recommendations for tanks holding petroleum oil are given in Chapter 9.

(b) The location of the tank in relation to its surroundings. A water tower is a conspicuous object, and a swimming pool with an underwater viewing gallery, or a pool constructed on the upper floor of a building, would normally all have to be 'bottle-tight', so that even slight seepage could not be tolerated, or special measures taken to mask it. These special measures may take the form of double walls and/or floors. On the other hand, the successful passing of the leakage test, set out in Clause 32 of Code of Practice BS 5337, would suffice for the vast majority of water and sewage tanks constructed on or in the ground.

A fundamental consideration in dealing with impermeability in concrete is that this material possesses a pore structure and in this respect is basically different to metals. This is why even the highest quality concrete is not gas- and vapour-tight. The capillary pore structure of concrete allows water under pressure to slowly penetrate the material. In high quality, well-compacted concrete, the rate of penetration is extremely slow. Permeability tests in the laboratory are difficult to carry out in such a way that the results will have practical application. Laboratory work is usually confined to low pressures of about 1·0–1·5-m head ($\frac{1}{10}$–$\frac{1}{6}$ atm).

It is perhaps relevant to note that the velocity of flow of ground water (rate of penetration) through a dense clay may not exceed 100 mm per year. High quality concrete is much less permeable than the densest clay.

The pressures in the vast majority of concrete liquid retaining structures are not likely to exceed about 10·0–12·0-m head (1·0–1·2 atm). If this is compared with a prestressed concrete pressure pipeline operating under, say, 150-m head, there need be no fear that the extremely slow passage of water through the floor or walls of a reservoir will have any detrimental effect whatever on the durability of the structure. This slow passage of moisture will not reduce the intense alkalinity of the concrete around the reinforcement (the alkalinity provided by the cement matrix is necessary for the effective prevention of corrosion of the steel).

Research carried out by the US Navy on hollow reinforced concrete spheres submerged at a depth of about 1000 m in the sea, showed that little if any seepage took place into the spheres.

The factors governing the permeability of concrete are very complicated and even today there are differences of opinion among concrete technologists on the relative importance of the many variables involved. Readers who wish to go into this subject in detail should refer to the bibliography at the end of the chapter.

However, most concrete engineers agree that the permeability of concrete depends on many factors including the following:

(a) The quality of the solid constituents, cement and aggregates.
(b) The quality and quantity of the cement paste and the pore size and distribution, which in turn depends on the quantity of cement in the mix, the water/cement ratio, and the degree of hydration of the cement.
(c) The bond developed between the paste and the aggregate.
(d) The degree of compaction of the concrete.
(e) The absence of cracking due to primary or secondary stresses.
(f) The adequacy of curing.

Of these six factors, the water/cement ratio probably exercises the greatest influence on permeability. With water/cement ratios exceeding 0·50, the coefficient of permeability increases very rapidly.

It has been shown by Powers and others in the US that the water/cement ratio and the time required for the blocking of the capillary pores are directly related. Generally, with a maximum water/cement ratio of 0·5 the capillary pores will be effectively blocked after about 15 days from the casting of the concrete. With a water/cement ratio exceeding 0·7, the capillary pores do not close at all.

The specification should lay down the requirements to obtain maximum impermeability of the concrete. There are differences of opinion among practising engineers as to what the specification should contain. The two main schools of thought are those who claim that the main clause should be a minimum cube strength, and those who recommend a minimum cement content and maximum water/cement ratio as the dominating requirements. The problem of relating cube strength to strength in the structure is referred to in Chapter 10. The authors recommend a minimum cement content, a maximum water/cement ratio, and a minimum cube strength which should be compatible with the cement content and water/cement ratio, and is not necessarily directly related to the design stresses in the concrete. In view of the fact that the water/cement ratio is very difficult to measure in concrete, some experienced engineers consider that this is best controlled by the requirement for slump, which usually has a tolerance of ±25 mm.

While the properties of the aggregates do affect strength, this is only likely to be significant with poor aggregates, or where high strengths, in excess of 45 N/mm^2, are required. For example, a rounded gravel will generally give a rather lower strength than a crushed rock, other factors being equal; but rounded gravels are not readily obtainable in the UK at the present time. Soft limestones, and sandstones and gravels from certain parts of Bedfordshire and Northamptonshire will not give high strength concrete and abrasion resistance. However, there is no real evidence to show that aggregates with relatively high absorption, even up to 8%, will give a concrete which is more permeable than one made with aggregates having absorption of 3% and below.

Carbonation

Recent work by P. E. Halstead on RILEM Committee 12 CRC, suggests that carbonation may play a more important role in the corrosion of steel in concrete than was previously suspected. The cause and effect of carbonation may be briefly described as follows:

(a) Portland cement concrete is highly alkaline due largely to the formation of calcium hydroxide.
(b) This alkalinity in concrete (in effect in its pore water) can be reduced by acid compounds in the atmosphere, particularly by carbon dioxide and sulphur dioxide. While carbon dioxide plays the major role in this process, which is known overall as 'carbonation', the effect of sulphur dioxide in reducing the pH is included. Carbonation is characterized by a sharp change in the pH value.
(c) Even fine cracks will admit air containing carbon dioxide and the sides of the crack become carbonated.
(d) Carbonated concrete does not provide the necessary protection (passivation) to the steel reinforcement.

Carbonation is a very slow process in dense impermeable concrete, and the rate is largely determined by

(a) the chemical reaction between the carbon dioxide and the calcium hydroxide in the pore water resulting in the formation of calcium carbonate ($CO_2 + Ca(OH)_2 = CaCO_3 + H_2O$);
(b) the rate of diffusion of the CO_2 inwards through the carbonated concrete. This diffusion and the passage of the water formed in the reaction given above takes place through the pore structure of the concrete.

Process (a) (chemical reaction) takes place very rapidly and therefore, in the long term, process (b) is the decisive one. The more water there is in the concrete (the higher the water/cement ratio), the more rapidly will the carbonation take place. In this context, 'water in the concrete' means the original amount of water in the mix; it should not be confused with subsequent saturation of the hardened concrete with water from outside. In fact there is a general tendency for drier concrete to carbonate more quickly than saturated concrete.

A comparatively simple test for carbonation can be carried out on a piece of concrete (broken rather than sawn), by the application of phenolphthalein; the carbonated layer remains virtually colourless, while the uncarbonated concrete turns the applied phenolphthalein bright pink.

Alkali–Aggregate Reaction

During 1977 a few cases of alkali–aggregate reaction were reported and confirmed on the mainland of the UK. During the past 30 or 40 years cases have occurred in the US, Denmark, Iceland, Germany, the Channel Islands, Canada, South Africa and New Zealand. A considerable amount of research has been carried out, and is continuing, into the various factors involved.

The reaction between the alkalis in Portland cement (mainly the sodium and potassium hydroxides) and a particular reactive form of silica found in certain aggregates, in the presence of water, can cause expansion and serious cracking in hardened concrete. When this reaction occurs within the concrete, an alkali silica gel is formed which is expansive in nature, and considerable forces are generated which can result in cracking. The formation of the gel and build-up of the expansive forces takes place very slowly and it is likely to be many years before signs of trouble appear.

A very special set of circumstances is required for this reaction to take place:

(a) The presence of a particular type of reactive silica in the aggregate. The proportion of this reactive silica must be within a certain narrow range which results in the maximum expansion. With too little or too high a proportion, the reaction will not take place.
(b) A Portland cement of relatively high alkali content, normally expressed as equivalent sodium oxide (Na_2O).
(c) A source of external water, sufficient to maintain the concrete in a wet condition.

The above conditions must all be present simultaneously for the reaction to occur and cause damage to the concrete. The damage is in the form of cracks, and the main body of the concrete does not suffer deterioration as occurs with sulphate attack. This has been proved by cores taken from badly cracked columns which gave equivalent cube strengths in the range 70–100 N/mm^2.

Recent work in Germany suggests that the total alkali in the concrete rather than in the cement alone may be significant. Therefore the addition of any admixture which may increase the total caustic alkali content of the concrete could be a factor in promoting alkali–aggregate reaction.

The difference between the reactive and non-reactive form of silica appears to be closely related to the crystalline structure.

The fact that the aggregate does contain a reactive form of silica does not necessarily mean that alkali–aggregate reaction will take place, as all the conditions mentioned above, namely (a), (b) and (c) must be present at the same time. If the alkali content of the cement does not exceed 0·6% as equivalent Na_2O, then reactivity is not likely to occur.

Some sulphate resisting Portland cements have relatively low alkali contents; the same applies to high alumina cement, which, according to Lea—*The Chemistry of Cement and Concrete*, p. 495, has an alkali content of 0·5% or less.

When it is found necessary to use aggregates from a previously unknown or untried source, it would be prudent to investigate whether they contain reactive silica, and if so whether, with the cement to be used and environmental conditions, reactivity is likely to occur. It is unfortunate that so far it has not been found possible to develop an effective test which will determine with certainty whether or not reactivity will take place.

Resistance to Aggressive Environments

General Considerations

As concrete forms about 99% of the structural materials with which this book is concerned, detailed consideration will be given to its use in a number of aggressive conditions commonly met in practice.

Aggression to concrete used for liquid retaining structures is likely to originate from one of the following causes:

(a) Aggressive compounds in the sub-soil and/or ground water.
(b) Aggressive chemicals in the air surrounding the structure.
(c) Aggressive liquid stored in the structure.
(d) Physical deterioration caused by freezing and thawing, and abrasion.

Aggressive Compounds in Sub-soil and Ground Water

The first important point is that chemicals in the dry state will not attack concrete; this applies particularly to sulphates as these are often found in the form of lenses in clay. The tolerable level of sulphate concentration in relatively dry, well drained soils is about four times the acceptable limit when the sulphates are in solution in ground water. In most soils it is the concentration of aggressive chemicals in the ground water which decides whether or not protective measures have to be taken.

There are two main categories of ground water, namely naturally occurring ground water, and ground water from industrial tips. Both types can be aggressive to concrete, but that from industrial tips is likely to be far more aggressive and difficult to deal with.

The range of chemicals found in industrial tips is much wider than in natural ground water. Drainage from these tips may contain sulphuric, hydrochloric, nitric, phosphoric and phenolic acids, ammonium compounds and sulphates.

When the report on the sub-soil investigations shows that the site is part of or adjacent to an industrial tip, or the analysis of the soil/ground water indicates the presence of industrial chemicals, expert advice should be sought.

All Portland cement concrete is vulnerable to acid attack as well as a number of industrial chemicals. The range of these chemicals is so vast that it is not practical to list them and make recommendations for protective measures for all cases. However, the authors feel it would be helpful to mention some of the chemical compounds which may be encountered in ground water and which are aggressive to Portland cement concrete, and this is done later in this chapter.

In what may be termed 'naturally occurring ground water', the acids most likely to be met are organic, such as humic acid derived from peat, and carbonic acid derived from dissolved carbon dioxide. Occasionally, water from marshy ground may contain sulphuric acid arising from the breakdown of sulphur compounds by bacterial action. This is another reason why a chemical analysis is important. The presence of sulphates in solution in ground waters, particularly those in clay soils, is much more common than acidity.

Acids, all types, in ground water. When the pH of the sub-soil and/or ground water is below the neutral point of 7·0 the water is acidic and is liable to attack concrete made with any type of Portland cement. The severity of the attack depends on a number of factors, the principal ones being

(a) the type and quantity of acid present;

Table 6.1

Concentration in solution	*pH Value*				
	HCl	*Acetic*	*Sulphurous*	*Carbonic*	*Boric*
N	0·1	2·4	—		
$\frac{N}{10} = 0{\cdot}1N$	1·1	2·9	4·9	3·8	5·2
$\frac{N}{100} = 1\% = 0{\cdot}01N$	2·0	3·4			

'N' is a normal solution, which is one containing the gram-equivalent weight of the solute in 1 litre of solution. For example, the gram-equivalent weight of HCl is (1 + 35·5) = 36·5.

(b) whether the acids are likely to be continuously renewed;
(c) the pressure and velocity of flow of ground water against the concrete;
(d) the cement content and impermeability of the concrete.

The pH by itself does not give any indication of the type or the amount of acid present. It is a measure of the intensity of the acidity, and can be considered as analogous to voltage of electric currents. Thus while pH measurements are useful as a rough guide, and clearly show the presence (or absence) of acids, they should, in most cases, be supplemented by chemical analysis to determine the nature and quantity of acid present. Table 6.1 shows variations of pH with some typical solutions of acids in various concentrations.

Table 6.2 is intended as a guide only, and each case should be treated as an individual problem.

It should be noted that the pH scale is logarithmic, so that a pH of 5·0 compared with a pH of 6·0 indicates an increase in acidity of ten times, but not a corresponding increase in aggressiveness.

Table 6.2. Acids in ground water

Probable Aggressiveness to Portland Cement Concrete			
Significant attack probably unlikely	*Slight attack probable*	*Appreciable attack probable*	*Severe attack probable*
pH 7·0–6·5	pH 6·5–5·0	pH 5·0–4·5	pH below 4·5

Table 6.2 should be read with the previous text and the following notes:

(a) Chemical analysis should preferably be carried out for most cases where the pH is below 6·5.
(b) Where the chemical analysis of the water suggests that slight attack only is probable, and provided this can be tolerated (as would normally be the case in a fairly massive reinforced concrete structure), the concrete should be made with either ordinary or rapid hardening Portland cement or Portland blastfurnace cement. The cement content should be not less than 330 kg/m^3, and the water/cement ratio should not exceed 0·50. The concrete must be fully compacted and properly cured.
(c) For conditions where the chemical analysis shows that appreciable attack is likely, the cement types should be as above, but the cement content should be increased to 360 kg/m^3 and the water/cement ratio should be reduced to 0·45. The use of plasticizers may be necessary to obtain the workability needed for full compaction of the concrete.
(d) Where severe attack is indicated by the chemical analysis, the concrete should comply with the previous conditions (c) for appreciable attack, but in addition it should be protected by substantial coating(s) of inert material, such as epoxide resin, polyester resin, or polyurethane, preferably reinforced with glass fibre membrane or strand.

The coating should not be less than 0·75 mm thick, and applied so as to eliminate pin holes. When practical, sheet material such as butyl rubber, PVC, chlorinated polyethylene, etc., can be used, but must be protected against physical damage. It is preferable for the sheeting to be bonded to the concrete, because if the membrane is damaged the area of concrete exposed to the aggressive liquid will be limited to the size of the hole in the sheeting. Care is therefore needed in the selection of the adhesive.

It should be noted here that sulphate resisting Portland cement is not considered to differ appreciably from ordinary Portland in its acid resisting qualities.

High alumina and supersulphated cements are resistant to a range of dilute acids, and supersulphated cement is also more resistant to sulphate attack than ordinary Portland cement. These two former cements can therefore be useful in foundation concrete which is subject to attack from the drainage from an industrial tip. However, expert advice should be sought before deciding to use either of them; such advice can be obtained from either the manufacturers of the cement or from the Building Research Establishment. At the time of going to press, supersulphated cement is no longer manufactured in the UK, but can be imported and HAC is at present not accepted for structural concrete under the Building Regulations.

Suggestions for the type of concrete which may be used under aggressive conditions have been given above. Under most conditions found in practice where naturally occurring ground water is involved, high quality, dense concrete is adequate. This can be supplemented by cut-off drainage, and backfilling with limestone or chalk to help neutralize the acidity; also an additional thickness of concrete to act as a 'sacrificial layer' is often a practical solution. The provision of 1000-gauge polyethylene sheets with bituminous adhesive (trade name 'Bitu-thene') bonded polyisobutylene (PIB) sheeting or bonded chlorinated polyethylene, along the bottom and carried up the sides to above top ground water level, can also provide reasonable protection.

The mass removal by mechanical excavators of industrial tip materials containing aggressive chemicals may be an economic and practical solution in some cases, particularly when it is combined with chalk or limestone backfilling and sub-soil drainage.

Sulphates in solution in ground water. Table 6.3 shows a range of ground conditions containing sulphates in solution in various concentrations and recommendations for type of concrete to be used. This is based on Building Research Establishment Digest No. 174—*Concrete in Sulphate-bearing Soils and Ground-waters*.

When Table 6.3 is used to help solve practical problems arising from sulphate concentrations, the following should also be noted:

(a) The sulphates most commonly found occurring naturally in ground water are
  (i) calcium sulphate (gypsum),
  (ii) magnesium sulphate, and
  (iii) sodium sulphate.
(b) Calcium sulphate is less soluble at normal temperatures than the other sulphates mentioned, and the solution is saturated at about 2000 mg/litre; this corresponds to about 1200 mg/litre of sulphur trioxide (SO_3). This means that under normal conditions of temperature, the concentration of calcium sulphate, expressed as sulphur trioxide, cannot exceed about 1200 mg/litre.

Table 6.3. Sulphates in soils and ground waters—Recommendations for concrete in contact with soil and ground water where any acids present are in too low a concentration to attack the concrete. (These recommendations apply to normal foundation concrete.)

Concentration of sulphates as SO_3		*Details of concrete mix*
In soil	*In ground water*	
0·2–0·5% (2 000–5 000 ppm)	300–1 200 ppm	Ordinary Portland cement, rapid hardening Portland cement, or Portland blastfurnace cement: 330 kg/m^3; water/cement ratio 0·50, or Sulphate resisting Portland cement: 300 kg/m^3; water/cement ratio 0·50, or Supersulphated cement: 310 kg/m^3, water/cement ratio 0·50
0·5–1·0% (5 000–10 000 ppm)	1 200–2 500 ppm	Sulphate resisting Portland cement: 330 kg/m^3; water/cement ratio 0·50, or Supersulphated cement: 330 kg/m^3; water/cement ratio 0·50
1·0–2·0% 10 000–20 000 ppm)	2 500–5 000 ppm	Sulphate resisting Portland cement or supersulphated cement: 360 kg/m^3; water/cement ratio 0·45
Over 2·0% (over 20 000 ppm)	Over 5 000 ppm	Sulphate resisting Portland cement: 400 kg/m^3; water/cement ratio 0·40 Adequate protective coatings of inert durable material may be required in addition

Notes

(a) Ammonium sulphate has not been mentioned as it is unusual to find this in natural ground water and soils, but it is quite usual to find this and other ammonium compounds in industrial tips. Generally, ammonium salts are very aggressive to Portland cement concrete and to reinforcement, and special precautions may have to be taken for which expert advice should be obtained.

(b) The first column of the table refers to 'relatively dry, well drained soils'. As previously stated, concrete is not attacked by sulphates in the dry state, and therefore it is assumed that the sulphates in this type of ground will in fact only go into solution very slowly. This is why the tolerable concentration in the soil is about four times that in the ground water.

(c) The sulphate concentration is shown as the equivalent of sulphur trioxide (SO_3). Sometimes sub-soil investigation reports show this as SO_4; conversion from SO_4 to SO_3 can be made as follows:

$$SO_3 = SO_4 \times \frac{80}{96} = 0{\cdot}83\, SO_4$$

(d) The concrete must be properly mixed, carefully placed and fully compacted. In cases where there is doubt about the degree of compaction likely to be obtained, the next higher class of concrete should be used.

(e) When a high proportion of the sulphate present is magnesium sulphate, it is advisable to use concrete in the next higher class.

(c) Sodium and magnesium sulphates are very soluble in water and therefore the concentration can be much higher than with $CaSO_4$.

(d) Magnesium sulphate is more aggressive to Portland cement concrete than the sulphates of sodium, calcium and potassium, when in equal concentrations. This means in practical terms that when the concentration is near the top end of a range, the recommendations for the next higher class should be adopted.

Magnesium sulphate has a more far-reaching action than other sulphates as it decomposes the hydrated calcium silicates in addition to reacting with the aluminates and calcium hydroxide.

Aggressive Chemicals in the Air Surrounding the Structure

It is only in very exceptional circumstances that concrete liquid retaining structures are likely to be attacked by aggressive chemicals in the air. Such cases may occur in chemical factories where a concrete structure is located above a vat or other container holding compounds giving off aggressive fumes.

The authors have known of cases where the fumes were acid, and caused severe attack on the concrete and steel reinforcement. Expensive remedial measures had to be taken and protective coatings applied to the outside of the concrete tanks.

It is relevant to remark that although chlorine is a poisonous and highly aggressive gas, it is used in the sterilization of 99% of the swimming pools in this country and chlorine compounds are consequently present in the air of swimming pool halls. However, the concentration is unlikely to exceed about 0·5 ppm (0·5 cm^3/m^3 of air). This will have no deteriorating effect whatever on good quality concrete. It will, nevertheless, increase the corrosiveness of moist, warm air towards ferrous metals and therefore all such metals should be given very high quality, durable protective coatings.

In sumps of pumping stations and sludge digestion tanks, conditions are sometimes met which result in the formation of sulphuric acid above top water level, by two-stage bacterial action. In such cases, special precautions are necessary, and reference can be made to the book *Concrete Structures: Repair, Waterproofing and Protection*, which deals with this subject in some detail.

Aggressive Liquids Stored in the Structure

General considerations. The range of chemicals used in modern industry is staggeringly large and to repeat what was said when considering aggressive ground water from industrial tips, each case has to be dealt with as a separate problem and given the most careful study.

Portland cement concrete is vulnerable to attack from all types of acids, sulphates in solution, and ammonium salts in solution, as well as a number of other chemicals. Brief notes on the more commonly used compounds are given below:

(a) *Acids.* The degree of attack depends on the chemical composition of the acid, its concentration and temperature and on the impermeability and cement content of the mix and the type of aggregate used. Reference should be made to the section on limestone aggregate concrete in this chapter (p. 146) where the effect of water from upland gathering grounds, containing organic acids and free carbon dioxide, is discussed in some detail.

(b) *Sulphates in solution.* Reference should be made to the detailed consideration given to aggressive ground water earlier in this chapter (p. 141). The only additional comment is that with stored liquids, the temperature may be appreciably above ambient, and while increase in temperature usually increases chemical activity, there are differences of opinion in the case of solutions of sulphates.

(c) *Brine.* This is usually a concentrated solution of sodium chloride (common salt). At normal temperature, this is not aggressive to concrete, but strong calcium chloride brine is likely to cause some deterioration in the concrete. It must also be realized that all brines are very aggressive to steel reinforcement and so maximum precautions must be taken to ensure impermeability and crack-free concrete.

(d) *Creosote.* This contains phenols and will slowly disintegrate concrete.

(e) *Lubricating oils.* If these contain fatty acids they are likely to attack concrete slowly; also they have high penetrating powers and if containing acids are potentially dangerous to reinforcement.

(f) *Petroleum oils.* These include petrol, paraffin and diesel oil. Generally these oils are not aggressive to concrete, provided they have a low fatty acid and sulphur content. However, they have high penetrating characteristics, and are likely to seep slowly through concrete unless an inert barrier is provided. The seepage is very slow and may show as dark stains and moist, oily patches on the outside of the container. Reference should also be made to Chapter 9, Section 9.5.

Fig. 6.1. *View of trade effluent settling tank showing protective epoxide resin coating. (Courtesy: Colebrand Ltd, London.)*

(g) *Sulphur (molten).* Liquid sulphur is not aggressive to concrete.

(h) *Caustic soda and other caustic alkalis.* Sodium hydroxide up to a concentration of about 10% is harmless to good quality concrete; concentrations above 10% are likely to cause slow disintegration. Increase in temperature of the solution will increase the speed of attack.

(i) *Sewage.* Domestic sewage is not aggressive to Portland cement concrete; septic sewage with a pH below 6·5 may attack concrete. Sewage containing trade wastes should always be carefully investigated and protection to the concrete may be required. Figure 6.1 shows a trade effluent tank protected by epoxide resin.

In certain circumstances sewage will become septic and hydrogen sulphide may be formed in significant quantities by anaerobic bacteria in the slimes and deposits below the water line in sewers and sumps of pumping stations. This gas (H_2S) rises into the airspace above the sewage and can be converted to sulphuric acid by aerobic bacteria in the presence of moisture. The sulphuric acid thus formed is very aggressive to Portland cement concrete and

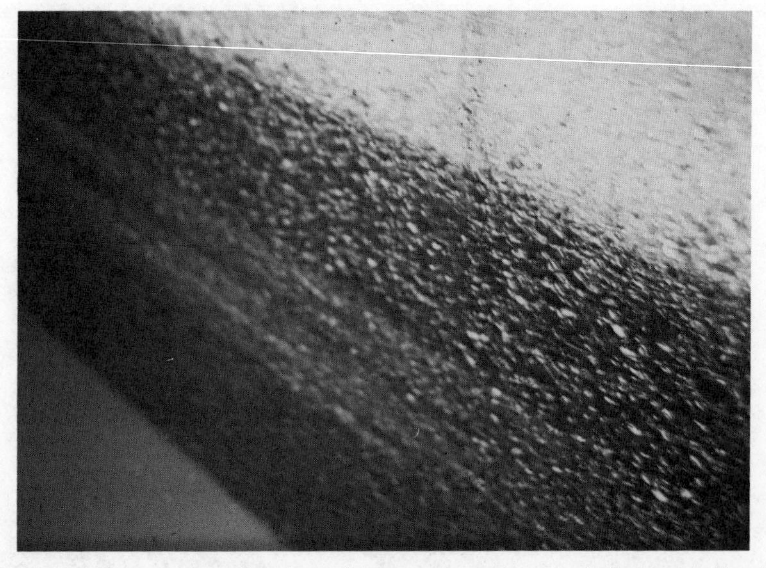

Fig. 6.2. *Close-up view of surface of concrete in a contact tank showing etching by aggressive water.*

when these conditions are anticipated, expert advice should be obtained.

(j) *Chemicals for water treatment*. Various chemicals are used for water treatment some of which are aggressive to Portland cement concrete; these include aluminium sulphate and ferrous sulphate. Both these compounds form acid solutions. The degree of attack depends on the concentrations and temperature. Figure 6.2 shows etching of the surface of the concrete walls of a contact tank after about seven years' service where ferrous sulphate was used.

For further information on this subject, reference should be made to the bibliography at the end of this chapter.

The Effect of Soft, Acid Waters on Portland Cement Concrete

Public water supplies which are derived from upland gathering grounds, such as moorland areas, usually have the following characteristics:

(a) Low pH values, in the range 4·0–6·5.
(b) Very low total dissolved solids.
(c) Low temporary and permanent hardness.

Fig. 6.3. *Close-up view of concrete surface etched by soft moorland water.*

(d) Low alkalinity (titration with methyl orange as indicator, expressed as calcium carbonate).
(e) Dissolved carbon dioxide.
(f) Dissolved organic (humic) acids.
(g) Occasionally, dissolved sulphuric acid, produced from the combustion of coal and fuel oil and other industrial processes, and from the oxidation of sulphur compounds in the peaty sub-soil.

All the above are variable throughout the year, particularly after heavy rain.

The above characteristics can cause the water to be aggressive to a greater or lesser degree to any type of Portland cement concrete. Figure 6.3 shows how this type of water will etch Portland cement concrete.

It is not possible in this book to consider in any detail all the above variables. The following is intended as a summary of the more important factors which should be taken into account when specifying concrete which will be in contact with this class of water.

The first essential is to obtain sufficient numbers of chemical analyses which will provide details of all the factors above, i.e. (a) to (g), with temporary and permanent hardness shown separately. Temporary hardness is hardness which is removed by boiling and consists mainly of calcium bicarbonate and magnesium bicarbonate. Permanent hardness consists principally of the sulphates of calcium, and magnesium and cannot be removed by boiling but only by chemical reaction.

It is essential to know the cause of the low pH, that is, whether it is due to organic or mineral acids, as the latter are likely to be much more aggressive towards Portland cement concrete. However, it is interesting to note that organic (peaty) acids are more likely to result in plumbo-solvency (the dissolution of lead in pipes and cisterns).

In making an estimate of the degree of attack which can be expected from a particular water (which does not contain mineral acids), the basic factor is the rate at which the concrete will be dissolved by the water. Concrete consists of a cement paste and aggregate. For the purpose of this discussion, the cement paste is considered to be mainly calcium hydroxide, and silica. The aggregate can be calcareous (limestone) or non-calcareous (igneous rocks, flint gravels, etc.). The surface of the concrete may be carbonated to a depth of a millimetre or so, so that calcium carbonate would then also be present in small quantities from the cement paste.

Calcium carbonate is very insoluble in nearly pure, neutral water (about 13 ppm); calcium hydroxide is comparatively very soluble (about 1200 ppm). All natural waters contain some carbon dioxide in solution, and this reacts with calcium carbonate to form calcium bicarbonate which is about four times as soluble as the carbonate. The high solubility of calcium hydroxide would suggest that the cement paste would be rapidly dissolved, but in fact this is not the case, because about 50% of the cement paste is silica and this protects the calcium hydroxide and greatly reduces the rate of attack.

The pH value itself has no direct relationship to the amount of acid present and therefore it cannot be used alone to predict the degree of attack by the water. The neutral point on the pH scale is 7·0; values below indicate acidity and values above alkalinity.

While with a pH in the range 7·0 to, say, 8·0, the water is alkaline, it can under certain conditions still be aggressive to Portland cement concrete, due to its potential in dissolving out the calcium hydroxide (lime). The dissolving power of such waters depends on the relationship of the total dissolved solids, the temperature, hardness expressed as $CaCO_3$, and the methyl orange alkalinity expressed as $CaCO_3$, and original pH of the water.

A very useful tool to determine whether or not a particular water is likely to dissolve calcium carbonate or to deposit it is the Langelier Index (Calcium Carbonate Saturation Index). This is a method of calculating from the methyl orange alkalinity, total dissolved solids, calcium hardness and water temperature, the pH at which the water will be just saturated with calcium carbonate. This is known as the saturation pH and is denoted by pH_s. The difference between the original (actual) pH of the water and the pH_s is the Langelier Index. A negative figure indicates that the water will tend to dissolve out lime, while a positive figure suggests that lime may be deposited; in other words, that the water will not attack the concrete.

The original equation as developed by Langelier is complicated to use, and so a number of water treatment specialists have developed simplified versions, using tables and/or graphs to solve the equation. A recent one is contained in a short paper by A. T. Palin (see the bibliography at the end of this chapter).

The following two examples illustrate the practical use of the Langelier Index to determine whether or not a water will be aggressive to Portland cement concrete. The calculations are based on Table 9.2, p. 287, of the text book *Water Treatment for Industrial and Other Uses* by E. Nordell. The following symbols and factors are used in this simplified equation:

Total dissolved solids (TDS) $= A$
Temperature (°C) $= B$
Calcium hardness as $CaCO_3$ $= C$
Methyl orange alkalinity as $CaCO_3$ $= D$

Example 1. Water 'X'

TDS: 100 ppm, $A = 0{\cdot}1$
Temperature: 13 °C, $B = 2{\cdot}3$
Calcium hardness as $CaCO_3$: 50 ppm, $C = 1{\cdot}3$
Methyl orange alkalinity as $CaCO_3$: 65 ppm, $D = 1{\cdot}8$
Original pH: 7·4

$$
\begin{aligned}
pH_s \text{ (saturation pH)} &= (9{\cdot}3 + A + B) - (C + D)\\
&= (9{\cdot}3 + 0{\cdot}1 + 2{\cdot}3) - (1{\cdot}3 + 1{\cdot}8)\\
&= 11{\cdot}7 - 3{\cdot}1 = 8{\cdot}6 = pH_s\\
\text{Saturation Index} &= 7{\cdot}4 - 8{\cdot}6 = -1{\cdot}2
\end{aligned}
$$

As this pH_s has a negative value, the water will tend to dissolve out $CaCO_3$ and is therefore potentially aggressive to Portland cement concrete.

Example 2. Water 'Y'

TDS: 500 ppm, $A = 0{\cdot}2$
Temperature: 14°C, $B = 2{\cdot}2$
Calcium hardness as $CaCO_3$: 230 ppm, $C = 2{\cdot}0$
Methyl orange alkalinity as $CaCO_3$: 280 ppm, $D = 2{\cdot}5$
Original pH = 7·8

$$pH_s \text{ (saturation pH)} = (9{\cdot}3 + 0{\cdot}2 + 2{\cdot}2) - (2{\cdot}0 + 2{\cdot}5)$$
$$= 11{\cdot}7 - 4{\cdot}5 = 7{\cdot}2 = pH_s$$
$$\text{Saturation Index} = 7{\cdot}8 - 7{\cdot}2 = +0{\cdot}6$$

The pH_s has a positive value, and therefore the water will tend to deposit calcium carbonate and will not be aggressive to Portland cement concrete.

Further information can be obtained by reference to the bibliography at the end of this chapter.

The preceding discussion leads naturally to consideration of the use of limestone aggregate in concrete for water retaining structures. As stated in Chapter 5, some authorities forbid the use of limestone when the structure will hold a soft, acid water. However, the cement paste and the limestone are basically similar chemically. Therefore if the cement paste and the limestone aggregate have approximately the same permeability, the water will attack both more or less equally. On the other hand if the aggregate is inert (a flint gravel, granite, etc.) to this class of water, then the whole of the attack from the water will be concentrated on the cement paste, resulting in quicker and deeper penetration of the concrete. The use of a high quality limestone as the aggregate for concrete is therefore generally recommended for structures holding or conveying a soft acid water.

There is no recognized test for limestone quality, which is admittedly a very variable rock; however, an examination of the quarry face will give a great deal of useful information, and reference can be made to BS 812—*Methods for Sampling and Testing of Mineral Aggregates, Sands and Fillers.*

For concrete in contact with these slightly aggressive waters, the authors recommend a minimum cement content of 360 kg/m^3 and a maximum water/cement ratio of 0·50; in addition the concrete should be fully compacted and properly cured. The authors have come across a few cases where concrete has been attacked by water of this class, and there appears to have been selective attack on the limestone aggregate. In these cases the general quality of the concrete in the structure was below that recommended above, and therefore definite conclusions have not been possible.

Sea Water

It is not often that liquid retaining structures are built in the sea, or contain sea or saline water, but it is felt that a few notes on the practical effect of sea water on concrete will be useful.

The total dissolved solids (salts) in Atlantic water is about 35 000 mg/litre (ppm). Of these about 2000 are sulphates (as SO_3) and 18 000 are chlorides.

The figure for sulphates is higher than that in class 1 soils and ground water (see Table 6.3). However, the authors have not seen any authentic reports of cases of chemical attack on good quality, dense, impermeable concrete by sea water around the coasts of the UK. The reason why a higher level of SO_3 concentration can be tolerated in sea water than in ground water has not been satisfactorily explained by research. It is generally thought that the other salts present may have an inhibiting effect on the reaction between the sulphates and the C_3A in the cement.

It should be noted that the above remarks refer to normal sea water. In some cases the water in estuaries is contaminated by sewage and trade wastes, and also if the tidal currents are obstructed, there may be a build-up of salts in solution, and sometimes aggressive organic compounds as well.

If sea water is the contained liquid, it may be at an elevated temperature and this increases chemical activity.

Because of the above, there are differences of opinion between engineers on whether or not sulphate resisting Portland cement should be used in structures built in the sea or containing sea water. The authors' experience is that unless there are special circumstances (sewage contamination, build-up of dissolved solids, high temperature, etc.), ordinary Portland cement is satisfactory except in areas such as the Persian Gulf, Red Sea, etc. The concrete must be high quality, dense and impermeable, with a minimum of 50-mm cover to all steel reinforcement.

A cement content of not less than 400 kg/m^3 and a maximum water/cement ratio of 0·45, special care in the positioning and fixing of reinforcement, and careful curing, are all necessary if deterioration is to be kept to a minimum. These conditions should be fulfilled even when sulphate resisting Portland cement is used. Further information can be obtained from the bibliography at the end of this chapter.

High alumina cement concrete has proved particularly resistant to sea water, but it is unlikely that a liquid retaining structure would be constructed entirely in HAC concrete. Reference should be made to Chapter 5 for information on this cement. It should be noted that HAC is not included in the list of cements permitted to be used for water retaining structures under Clause 21.1 of BS 5337. However, there would appear to be no technical or legal objection to the use of HAC rendering on a Portland cement base concrete.

Thermal swimming pools in spas usually contain naturally occurring spring water with a high concentration of dissolved salts. The chemical composition of these waters varies widely but generally analyses are readily available as this information is important for the 'cure'.

Each case should be considered as a separate problem and the solution lies in careful consideration of the chemical characteristics of the water and its temperature.

6.3. *The Use of Sea Water for Mixing Concrete*

In arid countries such as those on the Persian Gulf (Arabian Gulf) and the Red Sea, fresh water is extremely scarce and brackish water or even sea water may have to be used for mixing concrete. Questions then arise as to what effect this will have on the concrete.

If plain (unreinforced) concrete is used, the effect of the chlorides (mostly sodium chloride) in the sea water will be to accelerate the setting and hardening of the cement and to increase to some extent drying shrinkage. This is most undesirable in a hot climate as it will increase the risk of thermal cracking in the early life of the concrete. It will therefore be advisable to use a retarder in the mix to counteract the effect of the chlorides.

The effect of sulphates in the mixing water is unlikely to cause deterioration of the concrete because the amount of sulphate present is strictly limited and is not renewable. However, it is advisable to use sulphate resisting Portland cement even though the chlorides will tend to reduce its effectiveness (see Chapter 5).

In addition to the above, there may be appreciable efflorescence on the exposed surfaces of the concrete structure.

The position is more serious with reinforced concrete due to the probable corrosion of the reinforcement by the chlorides in the mixing water.

According to Lea—*The Chemistry of Cement and Concrete*, Atlantic water contains about 18 g/litre of chlorides, and Mediterranean water about 21·5 g/litre. The salinity of the sea in the Persian Gulf and Red Sea is higher, and therefore a figure of 22 g/litre (22 000 ppm) would be reasonable. This is 2·2 % by weight of the water. If the water/cement ratio of the mix is 0·45, then the chloride content of the concrete expressed as percentage by weight of cement would be $0{\cdot}45 \times 2{\cdot}2 = 1{\cdot}0\,\%$. It should be noted that this refers to chlorides and not sodium or calcium chloride. To convert sodium chloride content to chloride content, multiply by 0·61, and to convert chloride to equivalent sodium chloride, multiply by 1·64. These figures are arrived at from the chemical formula for sodium chloride and the atomic weights of sodium and chlorine, as follows:

Sodium chloride = NaCl; Na = 23·0, Cl = 35·5.

Therefore $23{\cdot}0 + 35{\cdot}5 = 58{\cdot}5$.

$$\text{Chloride content} = \frac{35{\cdot}5}{58{\cdot}5} = 0{\cdot}61 \qquad \text{and} \qquad \frac{58{\cdot}5}{35{\cdot}5} = 1{\cdot}64.$$

Some authorities consider that sodium chloride is rather less aggressive to ferrous metals than calcium chloride, but it would be unwise to take this into consideration.

In the case now being considered, additional chlorides are likely to be present as salt in the sand and possibly on the coarse aggregate. It can therefore be seen that the chloride content of the concrete would be over 1·0 % by weight of the cement, compared with the maximum permitted under the latest revision to CP110 of 0·35 %.

It is appreciated that there may well be cases where the only water available for mixing the concrete is sea water or brackish water and so the choice will lie between using this basically unsuitable mixing water or not proceeding with the project at all. In such circumstances it is unlikely that the project will be abandoned, and some reasonable solution (compromise) has to be found. The solution lies in taking the maximum practical precautions to protect the reinforcement from attack by the chloride ions. Some suggestions for this are given below.

(a) If brackish water is available, even though it is too saline for drinking, it is likely to contain considerably less salt than sea water. For example, in Kuwait, the brackish well water contained about 3800 ppm (mg/litre), of total dissolved solids (salts) compared with about 39 000 ppm in the sea water.

(b) If possible, the brackish water should be used to wash the sand and coarse aggregate.

(c) The cement content of the concrete mix should not be less than 400 kg/m^3, the water/cement ratio should not exceed 0·45, and the cover to reinforcement should not be less than 50 mm. In the opinion of the authors the cover should not exceed 60 mm owing to difficulty in controlling cracking.

Special care should be taken with compaction and curing.

(d) Wherever possible galvanized reinforcement should be used, and a Portland cement with a minimum chromate content of about 65 ppm. As an alternative to the use of such a cement (which in fact may not be available) chromates can be added to the galvanizing vats.

(e) As an alternative to (d), the reinforcement can be given two good coats of epoxide resin, after all rust and loose mill-scale has been removed. The hardened epoxide coating will reduce bond, but this can be partly offset by the use of high bond steel. An alternative to epoxide resin is a cement grout made with 2 parts sulphate resisting Portland cement to 1 part styrene butadiene latex emulsion. (See also Chapter 5.)

It is difficult to lay down definite limits for dissolved solids in mixing water, but a maximum of about 2000 ppm of the total ions of calcium, magnesium, sodium and potassium, as well as bicarbonate, sulphate, chloride and nitrate would be reasonable. The sulphate content should not generally exceed 1000 ppm (as SO_3), but this may have to be varied up or down according to the concentration of sulphates in the fine and coarse aggregates.

6.4. *Physical Aggression to Concrete Structures*

Physical aggression and consequent deterioration of concrete can arise from a number of causes. In the case of liquid retaining structures, these are likely to be limited to

(a) freezing and thawing on the outside of structures located in exposed positions in the northern part of the British Isles;

(b) abrasion from grit-laden water striking the concrete at medium or high velocity, or water relatively free of suspended solids with very high velocity.

Structures which are very exposed on high ground in the north of the country may suffer from surface scaling of the concrete due to frost action. This is caused by the penetration of the surface layers by moisture, followed by freezing temperatures. The consequent expansion of the water in the pores of the concrete may cause disruption and disintegration of the surface. Such conditions are rather rare in the UK but they do occur. Suitable preventive measures consist of the provision of dense, impermeable, high strength concrete, or the use of air entrained concrete. The authors feel that with concrete cast *in situ* on site, it may be difficult to achieve the necessary degree of compaction for complete protection. Hydraulically pressed precast products have proved their durability under such conditions, but it is not practical to think in terms of hydraulically pressed concrete for a water tower. In these circumstances, air entrained concrete provides the answer. Air entrained concrete will provide the necessary impermeability for a liquid retaining structure and at the same time will resist the disruptive effects of freeze–thaw. The air content should be $4\frac{1}{2}\% \pm 1\frac{1}{2}\%$. The air content must be carefully controlled by frequent use of an air meter.

The admixture must be thoroughly mixed in the gauging water before this is added to the cement and aggregate.

Regarding abrasion, this can take many forms, but the effect on a concrete liquid retaining structure is likely to be confined to specific and relatively small areas. Examples are sections of wall or floor opposite high velocity inlets, weir crests, and floors of channels carrying high velocity water, particularly if the water contains grit. These are all special cases which do not occur in the majority of liquid retaining structures.

While an appreciable amount of research has been carried out on the effect of very high velocity water on concrete, such as occurs on spillways where velocities of over 20 m/s are not uncommon, there is very little published information on the effect of velocities in the range 6–15 m/s. With very high velocities cavitation is a serious danger, but it must be realized that turbulence is a vital factor in the promotion of cavitation. No really satisfactory solution to this problem has yet been found. Advice is often sought on the effect of velocities in the ranges 4–6 m/s and 6–15 m/s.

In the lower range (4–6 m/s), the concrete should be high quality; a characteristic strength of 40 N/mm^2 (6000 lbf/in^2) at 28 days, a minimum cement content of 360 kg/m^3, and a maximum water/cement ratio of 0·5. The concrete as placed should be well compacted and properly cured. It is recommended that the aggregates should be carefully selected and the use of the softer limestones should not be permitted.

For velocities in the higher range (6–15 m/s), a specially high quality concrete is recommended, having a minimum cement content of 400 kg/m^3, a maximum water/cement ratio of 0·4, with heavy compaction and careful curing. In addition, special care should be taken to provide as smooth a surface as possible. High areas should be ground down and hollows or depressions or any honeycombing should be cut out and carefully filled with epoxide mortar. In order to obtain the necessary workability for full compaction with the low water/cement ratio recommended, a chloride-free plasticizer can be used.

However, where the flow is continuous over long periods and has a velocity exceeding 5 m/s, and/or contains an appreciable amount of grit, then additional and special precautions should be adopted. The nature of these special precautions will depend on the circumstances of each case, and in determining this, the following factors are important:

(a) The area of concrete involved.
(b) The possibility of carrying out periodic repairs.
(c) The velocity of the water and the amount of grit it contains.
(d) Changes in the shape of the containing structure.

Protective measures would include one of the following:

(a) A sacrificial layer, minimum 75 mm thick, of high strength concrete or gunite; maximum bond of this layer to the base concrete is essential.
(b) Three heavy coats of epoxide resin, applied after correct preparation of the base concrete, minimum 0·75 mm thick.
(c) A protective layer of epoxide resin mortar, containing an abrasive resistant aggregate, such as flint, or metallic aggregates. The thickness is likely to be in the range 10–25 mm.
(d) For horizontal surfaces only, a topping cast monolithically with the base concrete, consisting of Portland cement, finely divided iron, and a workability aid. The weight of iron may be as high as 45 kg/m^2, but even 5 kg/m^2 would effectively increase the abrasion resistance of the surface. In extreme cases it may be necessary to use steel plates fixed to the wall or floor.
(e) If there are changes in the shape of the tunnel, channel, etc., then expert advice should be sought as it is these places which are most vulnerable.

6.5. *Basic Principles of Concrete Mix Design*

General Considerations

The authors have found that sometimes there is uncertainty among engineers and site supervisory staff, of the principles involved in mix design.

It is therefore proposed to give some information in a simple form, on the important factors involved in the design of concrete mixes.

Mix design is defined in BS 2787—*Glossary of Terms for Concrete and Reinforced Concrete* as

> The calculations of the correct proportions of the constituents of concrete, taking into account such matters as the shape and grading of the aggregates and the workability and strength required in the concrete

The specifications for structural concrete for a liquid retaining structure should in most cases contain the following basic requirements:

(a) Minimum cement content, expressed in kg/m^3 (lb/yd^3) of compacted concrete.
(b) Maximum free water/cement ratio by weight (this assumes the aggregates are 'saturated surface dry').
(c) Workability, expressed in some standard form such as slump or compacting factor.
(d) Type of cement.
(e) Type of fine and coarse aggregate.
(f) Maximum size of the coarse aggregate.
(g) Minimum (characteristic) compressive strength of the concrete measured on cubes which have been prepared and tested in accordance with BS 1881—*Methods of Testing Concrete*. The 'characteristic' strength could be compatible with requirements (a) to (f), but particularly (a), (b) and (c). Because of the relatively high cement content and low water/cement ratio, the specified characteristic strength is likely to be higher than is required by the design assumptions alone.
(h) Admixtures, none, unless approved in writing by the engineer.

Concrete mix design is a subject on which a number of complete books have been written, as well as numerous papers and some of these are listed in the bibliography.

Mix Design

Turning to the problems of actual mix design, the following are the most important of the many factors (except the actual mixing) which decide the quality, used in its widest sense, of the concrete as it leaves the mixer.

(a) Type and quantity of the cement in the mix.
(b) Water/cement ratio.

(c) Type, grading, maximum size, shape and surface texture of the coarse aggregate.
(d) Type, grading, shape and surface texture of the fine aggregate.

The mix must be designed so that it can be fully compacted with the means available and under the conditions existing on the site. There must be no segregation between the mixer and the final resting place of the concrete; this is of special importance in pumping.

To enable concrete to be fully compacted the mix must be 'workable'. Although this is such an important matter, concrete technologists say that there is no really satisfactory test for workability. However, the slump test and compacting factor test are widely used and appear to work well in practice. The slump test is simple and easy to carry out, and is acceptable for the majority of projects. Compacting factor apparatus is only found on large construction sites where there is continuous, high-grade supervision, and in testing laboratories. A third method of measuring workability is the Vee Bee consistometer, which is used in laboratories.

The chief factor influencing workability is the water/cement ratio; with other things equal, the more water there is in the mix the more workable the concrete will be. The higher the water/cement ratio the lower the strength of the concrete.

It is emphasized that the recommendations given in this book for workability (expressed as 'slump') are for guidance only. The site supervising staff must have the authority to vary the slump if they consider this is necessary to prevent segregation, improve 'pumpability' or assist compaction.

The presence of closely spaced reinforcing bars, and the fixing of water bars in comparatively thin walls, will interfere with compaction, particularly if the mix is stiff and does not 'flow' easily under the action of the poker vibrators. In such cases, immediate action must be taken to improve the workability. To persist with a too-low slump concrete is almost certain to result in under compaction and consequent leakage. In specially difficult situations, consideration can be given to the use of super-plasticizers, which are briefly described in Chapter 5.

Some detailed comments on the four previously mentioned factors which affect the quality of the concrete mix are:

(a) The type and quantity of cement in the mix. For the purpose of this section it is assumed that only Portland cement (ordinary, rapid hardening, or sulphate resisting) is used.

With the water/cement ratio, and the aggregate type (grading and other characteristics) constant, the workability will increase with increase in cement content. An increase in cement content, within certain limits, will also improve cohesiveness, and increase strength and durability. Where very high strengths are required (characteristic strengths above 55 N/mm^2 (7500 lbf/in^2), it may be necessary to specially select the cement and to do the same with the aggregates, as well as maintaining a low water/cement ratio, probably 0·40 or lower.

(b) Water/Cement Ratio. As the water/cement ratio increases, workability increases (the other factors remaining constant), and strength and durability decreases.

(c) and (d) Characteristics of the fine and coarse aggregates.

(i) Type of aggregate. This will affect the shape and surface texture: for example, artificial lightweight aggregates of expanded clay will be different to crushed rock or gravel.

(ii) Maximum size of aggregate. As the maximum size of the coarse aggregate increases up to about 50 mm (2 in), the workability will also increase, other factors remaining constant. Thus with two mixes having the same cement content, water/cement ratio, and aggregate type, the mix with 40 mm ($1\frac{1}{2}$ in) coarse aggregate would be more workable than the one with the 20 mm ($\frac{3}{4}$ in) aggregate. It should be noted that when using a small size aggregate, such as 10 mm ($\frac{3}{8}$ in) gravel, more cement, more water, and a higher proportion of fine aggregate are needed in the mix to give the same workability.

(iii) Shape and surface texture of aggregate particles. Rounded aggregates with a smooth surface, which have a low absorption, will result in a more workable mix than say a crushed rock, but may give a concrete with a lower strength. Impermeability and resistance to abrasion are not likely to be affected. It is becoming increasingly difficult to obtain rounded aggregates for concreting in the UK.

(iv) Grading of aggregates. Generally the grading of aggregates has a greater effect upon workability than shape and surface texture. Most concretes used for structural purposes have what is known as continuous grading. Gap-graded aggregates can also make satisfactory concrete, but to obtain consistent results, close supervision is required, and compaction by vibration is essential. With a gap-graded aggregate, one part of the grading curve is horizontal. The effect of the 'gap' on the resulting concrete is largely governed by its location on the grading curve.

Crushed rock fines generally contain a higher percentage of fine material passing the 100 sieve than is the case with a natural sand, and this tends to increase the water demand of the mix, resulting in lower strength.

(v) Other factors. The percentage of sand (fine aggregate) in relation to the total weight of aggregate is also important. The larger the maximum size of the coarse aggregate, the lower the percentage of fine aggregate required for a given workability, with a constant cement content. The following shows the approximate percentage of sand required with different sizes of coarse aggregate:

10 mm ($\frac{3}{8}$ in) 50–55 % sand
20 mm ($\frac{3}{4}$ in) 30–45 % sand
40 mm ($1\frac{1}{2}$ in) 25–35 % sand

It is emphasized that the above are average figures. To maintain the same workability, the sand content may have to be increased if the cement content is reduced.

The variations in constituent materials of concrete mixes with different workabilities, etc., is well illustrated in the Cement and Concrete Association's advisory booklet—*Concrete Constituents and Mix Proportions* and the more recently published DoE booklet—*Design of Normal Concrete Mixes*.

It must be remembered that both cement and aggregates are fairly 'crude' materials, and even though they comply with national standards, they are distinctly variable, and these variations affect the characteristics of the concrete. This applies particularly to strength and workability. To obtain a specified slump with a specified maximum water/cement ratio may require the concrete producer to increase the specified minimum cement content by, say, 100 kg/m^3. Therefore early discussions between the engineer, the contractor and the supplier of the concrete are essential if concrete of the required quality is to be supplied consistently, and disputes avoided.

6.6. *Calculations for the Yield of Concrete Mixes*

For liquid retaining structures, all batching should be by weight and therefore volume batching is not included here.

Concrete Batching by Weight

For all practical purposes it can be assumed that 1 m^3 of compacted concrete weighs 2385 kg (1 yd^3 weighs 4000 lb). Weights can be assumed to be absolute, and therefore the weight of concrete will equal the total weight of constituent materials.

Example: Find the volume of concrete produced per batch when using the following batch weights:

Cement:	50 kg
Sand:	90 kg
Coarse aggregate:	160 kg
Water/cement ratio:	0·50

Water: (i.e. calculated weight of water in the sand plus the amount added at the mixer):

$$0{\cdot}50 \times 50 = 25\,\text{kg}$$

Total weight: 325 kg = weight of concrete per batch

Volume of concrete

$$= \frac{325}{2385} = \left(\frac{\text{weight}}{\text{density}}\right) = 0{\cdot}137\,\text{m}^3 = 137\,\text{litre}$$

For most constructional purposes, it is sufficiently accurate if the figure of 2385 kg is rounded off to 2400 kg.

In imperial measure, using the above and approximate conversion factors:

Cement:	112 lb
Sand:	200 lb
Aggregate:	360 lb
Water:	56 lb
	728 lb

$$\text{Volume of concrete per batch} = \frac{728}{4000} = 0{\cdot}182\,\text{yd}^3 = 4{\cdot}9\,\text{ft}^3$$

6.7. *Calculations for the Yield of Mortar Mixes*

Mortars for Screeds

Batching cement and sand for mortar for rendering and screeds should be by weight because it is only in this way that uniform quality of the mortar can be achieved. Unfortunately, although contractors and specifiers recognize the need to batch concrete by weight they seem to make little or no attempt to ensure that mortar is batched by weight; the exception is the suppliers of pre-mixed (ready-mixed) mortars.

There is no standard density for mortars as these are not normally specified by strength, and compaction generally leaves much to be desired. Sometimes, however, an attempt is made to specify a strength for mortar. Cubes made in the laboratory with Leighton Buzzard sand and mix proportions by weight of 1:3 and water/cement ratio of 0·40 are likely to give very much higher strengths than will be achieved on site.

Laboratory strengths may be

1 day	14 N/mm^2 (2000 lb/in^2)
3 days	30 N/mm^2 (4000 lb/in^2)
7 days	40 N/mm^2 (6000 lb/in^2)
28 days	50 N/mm^2 (7000 lb/in^2)

To calculate the yield of a mortar mix, one can either use the absolute volume method, or assume a density of, say, 1940 kg/m^3 (120 lb/ft^3). The absolute volume method gives rather a higher density than 1940 kg/m^3 so that the calculated yield in terms of volume is lower than when a more realistic density figure is assumed. The authors prefer to assume a density figure of 1940 kg/m^3 (120 lb/yd^3).

The following example shows this simple calculation:

Example: Calculate the yield of a mortar mix having the following proportions by weight:

Cement: 50 kg
Sand: 150 kg
Water: 25 kg (this is the total water in the batch with a water/cement ratio of 0·50)

Total: 225 kg

$$\text{Volume of mortar: } \frac{225}{1940} = 0{\cdot}116\,m^3 = 116\,\text{litre}$$

Using the absolute volume method:
Specific gravity of cement: 3·12
Specific gravity of sand: 2·65
Specific gravity of water: 1·00

Density of water: $1000\,kg/m^3 = 1000\,g/litre$

$$\frac{50}{3{\cdot}12 \times 1000} + \frac{150}{2{\cdot}65 \times 1000} + \frac{25}{1{\cdot}0 \times 1000}$$

$$= 0{\cdot}0975\,m^3 = 97{\cdot}5\,litre$$

This gives a density of $2300\,kg/m^3$ ($143\,lb/ft^3$)

Neither method of calculation takes into account loss and wastage of materials. It is not possible to give a fixed percentage for this as the wastage and loss depends on how the cement and aggregates are stored, batched and mixed. The quantities involved are also relevant; for example, a job requiring 500 kg ($\frac{1}{2}$ ton) of sand may well involve a wastage/loss of 50–75 kg (112–160 lb), but it is very doubtful if on a job using 100 tonnes of sand in a more or less continuous operation that a wastage/loss of 10–12·5 tonnes would be accepted.

Another method, which is quicker and sufficiently accurate for many purposes, is used when the mortar is batched by volume which, as previously stated, is still the usual method. The volume of mortar is assumed to be equal to the volume of the sand, and the cement and water are assumed to fill the voids in the sand.

Example: Find the quantities of cement and sand required for a screed with an area of $1000\,m^2$, 50 mm thick and the mortar mix is 1 part cement to 3 parts sand by volume.

Volume of mortar: $1000 \times \frac{50}{1000} = 50\,m^3$

Volume of sand: $50\,m^3$

Volume of cement: $\frac{50}{3} = 16{\cdot}67\,m^3$

Bulk density of sand: $1600\,kg/m^3$

Weight of sand: $50 \times \frac{1600}{1000} = 80$ tonnes

Bulk density of cement: $1400\,kg/m^3$

Weight of cement: $16{\cdot}67 \times \frac{1400}{1000} = 23{\cdot}4$ tonnes

The above does not take into account any loss of material on site.

Bibliography

American Concrete Institute. Guide for the protection of concrete against chemical attack by means of coatings and other corrosion-resistant materials. Report by ACI Committee 515. *J. ACI*, December 1966, pp. 1305–1390.

American Concrete Institute. *Manual of Concrete Practice. Part 3—Products and Processes*, 1972, pp. 515-23 to 515-73.

Anon. Sea tests on concrete spheres. *Concrete*, December 1977, 35 pp.

Australasian Corrosion Association. *Corrosion Assessment and Control.* Conference 14, Brisbane, Australia, November 1973, Twenty-six papers.

British Standards Institution. *Commentary on Corrosion of Bi-metallic Contacts and its Alleviation.* PD 6484, London, 1979.

Building Research Establishment. *The Durability of Reinforced Concrete in Sea Waters; Twentieth Report of the Sea Action Committee of the Institution of Civil Engineers National Building Studies.* Research Paper No. 30, 1960, HMSO, London, 42 pp.

Building Research Establishment. *Concrete in Sulphate-Bearing Soils and Ground-Waters.* Digest 174 (Second Series), 1972, HMSO, London, 6 pp.

Building Research Establishment. *Carbonation of Concrete Made with Dense Natural Aggregates.* Information Sheet 15/14/78, August 1978, 3 pp.

Canadian Portland Cement Association. *Design and Control of Concrete Mixtures.* Portland Cement Association, Skokie, USA, 1978, 131 pp.

Everett, L. H. and Treadaway, K. W. J. *The Use of Galvanized Steel Reinforcement in Building.* Current Paper CP 3/70, Building Research Establishment, January 1970, 10 pp.

Fulton, F. S. Interpretation of water analyses. *Concrete Technology*, Fifth Edition. The Portland Cement Institute, Johannesburg, 1977, 421 pp.

Harrison, T. A. *Tables of Minimum Striking Times for Soffit and Vertical Formwork.* CIRIA Report No. 67, October 1977, 24 pp.

Harrison, T. A. *Formwork Striking Times—Methods of Assessment.* CIRIA Report No. 73, October 1977, 38 pp.

Halstead, P. E. Corrosion of reinforcement in concrete. *Materials Structures*, **9**(51), May–June, 1976, pp. 187–206.

Houston, J. T., Atimtay, T. and Ferguson, P. M. *Corrosion of Reinforcing Steel Embedded in Structural Concrete.* Research Report 112-1F, Centre for Highway Research, University of Texas, Austin, USA, March 1972, 131 pp.

Langelier, W. F. The analytical control of anti-corrosion water treatment. *J. American Water Works Assoc.*, **28**(10), October 1936, pp. 1500–1521.

Lea, F. M. *The Chemistry of Cement and Concrete*, Second Edition (Revised). Edward Arnold Ltd, London, 1970, 637 pp.

Murphy, F. G. *The Effect of Initial Rusting on the Bond Performance of Reinforcement.* CIRIA Report No. 71, November 1977, 28 pp.

Naval Civil Engineering Laboratory. *Behaviour of 66-inch Concrete Spheres under Short-Term and Long-Term Hydrostatic Loading.* National Technical Information Service, Springfield Va., USA, 1972, 86 pp.

Nordell, E. *Water Treatment for Industrial and Other Uses*, Second Edition. Reinhold, New York, and Chapman and Hall, London, 1961, 598 pp.

Palin, A. T. A simplified approach to balanced water calculations. *J. Inst. Water Engrs Scientists*, **32**(5), September 1978, pp. 421–422.

Palmer, D. *Alkali–Aggregate (Silica) Reaction in Concrete*. Code 45.033, Cement and Concrete Association, London, December 1977, 8 pp.

Perkins, P. H. *The Use of Concrete in Sulphate Bearing Ground and Ground Water of Low pH*. ADS 30, Cement and Concrete Association, London, April 1976, 16 pp.

Perkins, P. H. *Concrete Structures: Repair, Waterproofing and Protection*. Applied Science Publishers, London, 1977, 302 pp.

Perkins, P. H. The durability of concrete in underground pipelines. *Paper G1, Second International Conference on Internal and External Protection of Pipes, Canterbury*, September 1977, 11 pp.

Perkins, P. H. *Swimming Pools*, Second Edition. Applied Science Publishers, London, 1978, 398 pp.

Phillips, E. *Survey of Corrosion of Prestressing Steel in Concrete Water Retaining Structures*. Department of the Environment and Conservation, Australian Resources Council Research Project 71/32, Canberra, 1975, 143 pp.

Powers, T. C., Copeland, L. E. and Mann, H. M. Capillary continuity or discontinuity in cement pastes. *J. Portland Cement Assoc. Res. Dev. Lab.*, May 1959, pp. 38–45.

RILEM. Corrosion of reinforcement in concrete. *Materials and Structures*, **9**(51), May/June 1976, pp. 187–206.

RILEM–FIP–IABSE Committee. *Corrosion Problems with Prestressed Concrete*. Fifth Congress of FIP, Paris 1966, 6 pp.

Shacklock, B. W. *Concrete Constituents and Mix Proportions*. Cement and Concrete Association, London, 1974, 102 pp.

Terzaghi, K. and Peck, R. B. *Soil Mechanics in Engineering Practice*, Third Edition. John Wiley & Sons, New York, and Chapman and Hall, London, 1968, 752 pp.

Teychenne, D. C., Franklin, R. E. and Erntroy, H. C. *Design of Normal Concrete Mixes*. HMSO, London, 1975, 30 pp.

Tomlinson, M. J. *Foundation Design and Construction*, Third Edition. Pitman Publishing, London, 1975, 785 pp.

Treadaway, K. W. J. and Russell, A. D. *Inhibition of the Corrosion of Steel in Concrete*. Current Paper CP 82/68, Building Research Establishment, December 1968, 5 pp.

CHAPTER 7

BASIC CONSTRUCTION TECHNIQUES

7.1. *Introduction*

This chapter sets out to discuss a number of important subjects which are applicable to most types of concrete liquid retaining structures. These matters are concerned with the use of concrete as a material and the authors consider them of fundamental importance to both the designer and constructor.

It is not possible to draw a clear dividing line between design, specification and construction. Designers should possess a working knowledge of how the structure will be built and the problems the contractor is likely to face on site. Equally necessary, the contractor should know the basis for the design and reasons for the specification requirements.

There is a detailed discussion on thermal contraction in the concrete of walls, floors and roofs, due to the heat of hydration of the cement, with consequent expansion followed by contraction as the concrete cools. This discussion is concerned with the practical aspects of the problem. The revised Code of Practice, BS 5337, includes recommendations for design which are intended to ensure that this type of cracking is 'controlled' in as much as the crack spacing and crack width are predetermined by using the design criteria in the Code. General comments on the Code are given in Chapter 11. Examples of design using the methods in the Code are given in Part 1 of this book.

7.2. *Concrete Construction in Cold Weather*

General Considerations

The principles of winter building in general and winter concreting in particular are much better known and understood in the UK than was the case even 10 years ago.

The information and recommendations given here are intended to bring out important principles. Readers who wish to go into the subject in detail should refer to the bibliography at the end of this chapter.

There are a number of important factors involved in using Portland cement concrete when air temperatures are near freezing point and these include the following:

(a) When the temperature of setting and maturing concrete is lowered the chemical reactions involved are slowed down and therefore the rate of gain of strength decreases. As freezing point in the mass of concrete is approached the hardening process practically ceases altogether. The temperature at which hardening ceases is, for the purpose of maturity calculations, assumed to be −10 °C.

(b) However, if the concrete is not saturated with water, and if it has reached a compressive strength of about 3 N/mm^2 (450 lbf/in^2) and has a reasonable cement content (300 kg/m^3), then even if the concrete does freeze, it is unlikely that permanent damage will result. When the temperature of the concrete rises again the concrete will thaw out and the maturing process (hardening) will re-commence and continue at a rate proportional to the temperature of the concrete.

(c) It is the temperature of the concrete which is the key factor, although this is of course directly influenced by the ambient temperature and any thermal insulation which may be used to protect the concrete. Even if the compressive strength of the concrete is less than 3 N/mm^2, as mentioned in (b) above, no harm will occur to the concrete provided its temperature does not reach freezing point. If freezing point is reached in the concrete with a lower strength, then it is likely that permanent damage will be done and the concrete will have to be removed.

(d) The precautions to be taken to prevent damage to the concrete by low temperatures must all be directed towards maintaining the temperature of the concrete as high as practical. However, the temperature of the heated concrete at the time of placing should not generally exceed about 30 °C (86 °F).

Basic Precautions to be Taken

The following simple rules, if properly applied, will enable concreting to proceed even in the most severe weather likely to be experienced in the UK:

(a) Frozen aggregates and icy water must not be used.

(b) Concrete must not be placed in frozen forms nor on frozen ground.

(c) Concrete should contain not less than 300 kg of Portland cement per cubic metre of compacted concrete (500 lb/yd^3).

(d) Wet curing techniques should not be used.

(e) The concrete temperature at the time of placing should not be lower than about 6 °C (43 °F). This can be achieved by ordering heated concrete from a ready-mixed concrete plant which is equipped to produce this type of concrete. If the concrete is mixed on site the aggregate stock piles must be covered with tarpaulins and/or steam jets applied before the aggregates are used; heated water should also be used for mixing, but the temperature should not exceed about 60 °C (140 °F).

(f) Steps should be taken to ensure that there is a rapid gain in strength in the concrete after it is placed and compacted. This can be achieved by increasing the cement content by, say, $1\frac{1}{2}$ bags (75 kg) per cubic metre of concrete; an alternative is to use rapid hardening Portland cement, or an accelerating admixture which is not based on calcium chloride.

(g) Finally, the concrete must be well insulated against loss of heat immediately the finishing process is complete.

It may not be necessary to adopt all the measures listed above and selection will depend on the severity of the weather and the practical possibility of carrying them out. Regarding accelerating admixtures, reference has been made to the use of these compounds in Chapters 5 and 6.  Admixtures which are not based on calcium chloride are now available, but except in most unusual circumstances, they are unlikely to be used in the construction of new liquid retaining structures in the UK, but they may be useful for emergency repairs.

When considering what measures should be adopted to protect freshly placed concrete from the effect of very low temperature, the following factors should be taken into account:

(a) The exposure conditions of the site and the structural units being concreted.

(b) The surface area to volume ratio. In this case, the surface refers to the exposed surface of the concrete not protected by timber formwork. The higher the ratio the more vulnerable the concrete will be to excessive heat loss. For example, floor slabs are appreciably more vulnerable than walls. Toppings and screeds on roofs are very vulnerable indeed and in some cases this type of work should be suspended until the weather improves.

The following should also be noted:

(i) The grouting of prestressing ducts should in most cases be suspended; this also applies to pressure grouting of defective concrete.
(ii) It is not advisable to carry out rendering in very cold weather due to excessive heat loss.
(iii) The building of external blockwork and brickwork should preferably be suspended when the ambient day air temperature is near freezing as the mortar in the joints is very vulnerable to the effect of low temperature.

7.3. *Concrete Construction in Hot Weather*

General Considerations

It may be thought that in a temperate climate such as exists in the UK there would be no problems arising from concreting in hot summer weather. However, difficulties can be experienced with the premature stiffening of concrete which makes compaction very difficult; there is also the increased risk of thermal contraction cracking, and plastic cracking. The consequences of neglect in proper curing are likely to be more serious in hot weather than in cold and cool weather.

The use of heat, e.g. heated concrete, to speed up the chemical reaction of the mixing water and the cement has been mentioned in Section 7.2 on concreting in cold weather. The effect of heat on this reaction is the same irrespective of its source. In summer the sun is the source of this external heat. During hot weather the mixing water, aggregates and formwork are all warm or hot. The higher the initial temperature of the concrete the more quickly will the reaction proceed which in turn generates heat more rapidly, resulting in a 'snowball' effect. As an example of this, concrete placed at, say, 10 °C (50 °F) may reach a temperature of 30 °C (86 °F) in, say, 24 h; if the same mix of concrete were placed under exactly the same external conditions but with an initial temperature of, say, 20 °C (58 °F), the maximum temperature reached may be 55 °C (130 °F) after about 20 h.

The rapid temperature rise in the hydrating mass of concrete will be increased with increase in cement content, but this rise will be modified if the exposed surface area to volume ratio is increased.

Also the higher the heat of hydration of the cement used, the higher the maximum temperature reached in the concrete—other things being equal.

Recommended Procedure

It is recommended that the following matters be given consideration when casting concrete in hot weather:

(a) When a cement content in excess of 330 kg/m^3 (550 lb/yd^3) is specified, it is worth while to investigate the possibility of reducing the amount of cement, but keeping in mind, strength, durability and impermeability.

(b) The aggregate stock piles can be sprayed with water as this has a significant effect on lowering the temperature of the aggregate. The reason for this is that the specific heat of water is 1·0 and the latent heat (conversion from liquid to gaseous state) is 540 000 cal/litre.

(c) The mixing water should be as cool as possible. Water drawn direct from a public supply main is usually cool, but if it is kept in a static water tank on site it may become warm or even hot. A static water tank can be insulated at low cost and two coats of white-wash are very effective in reflecting sun heat.

(d) While the temperature of the cement has only a marginal effect on the temperature of the mixed concrete it is worth while to maintain a reasonable stock so that it is not necessary to use fresh cement immediately it is delivered into the silo.

(e) The use of a retarder in the mix can be very useful and in hot climates it is likely to be essential. Trial mixes and strict control of the dosage are necessary.

(f) Careful curing is of the utmost importance, and it may be necessary, even in temperate climates, to use sun covers (tentage) to protect floor and roof slabs.

Factors Affecting the Temperature of Freshly Mixed Concrete

The following formula which is contained in an unpublished Cement and Concrete Association document shows the effect of the temperature of the main constituents of a concrete mix on the temperature of the concrete as it comes from the mixer.

$$T = \frac{0{\cdot}22(T_a W_a + T_c W_c) + (T_f W_f + T_m W_m)}{0{\cdot}22(W_a + W_c) + (W_f + W_m)}$$

where T = temperature of fresh concrete
T_a = temperature of aggregates
T_c = temperature of cement
T_f = temperature of free moisture in the aggregates
T_m = temperature of mixing water
W_a = weight of aggregates
W_c = weight of cement
W_f = weight of free moisture in the aggregates
W_m = weight of mixing water

In the above formula it is assumed that the specific heat of cement and aggregates is 0·22. This is sufficiently accurate in view of the obvious fact that the temperatures are unlikely to be measured simultaneously, and that as soon as water is added to the cement chemical action starts which generates heat, thus raising the temperature while mixing is taking place.

It was pointed out to the authors that on tests carried out by the Royal Engineers in Cyprus in 1973, during the construction of a swimming pool, the above formula gave results very close to the measured temperature of the concrete.

7.4. *Plastic Cracking*

There are two types of plastic cracking, plastic shrinkage cracking and plastic settlement cracking; the former is much more common.

Plastic Shrinkage Cracking

This type of cracking may occur on the surface of floor and roof slabs while the concrete is still plastic. Investigations by various authorities have shown that the principal cause of plastic cracking is a rapid evaporation of moisture (drying out) from the surface of the concrete. When the rate of evaporation exceeds the rate at which water rises to the surface (known as 'bleeding'), plastic cracking is very likely to result.

The rate at which the water in the mixed concrete reaches the surface and the total quantity involved depends on many factors, all of which are not yet completely understood, but the following are known to play an important part in this phenomenon:

(a) Grading, moisture content, absorption, and type of aggregate used.
(b) Total quantity of water in the mix.
(c) Cement content.
(d) Thickness of the concrete slab.
(e) Characteristics of any admixtures used.
(f) Degree of compaction obtained and therefore the density of the compacted concrete.
(g) Whether the formwork (or sub-base) on which the concrete was placed was dry or wet.

The rate of evaporation from the surface will also depend on a number of factors which are much better understood and these are

(a) relative humidity;
(b) temperature of the concrete;
(c) temperature of the air;
(d) wind velocity;
(e) degree of exposure of the surface of the slab to the sun and the wind.

It has been stated that if the relative humidity is reduced from 90% to 50% and all other factors remain constant, the rate of evaporation will be increased five times. With a reduction from 90% to 40%, the rate of evaporation will be increased nine times. If the concrete and air temperatures are raised from 10 °C to 20 °C (50 °F to 70 °F), the rate of evaporation will be doubled. However, reference to meteorological records will show that in the southeast of England the relative humidity is 95% and over for about 25% of the year and in the southwest of England for over 33% of the year.

Plastic shrinkage cracking shows itself as fine cracks which are usually straight and short. They are often transverse in direction and very rarely extend to the slab edge. The cracks are sometimes parallel to each other, and the spacing can vary from a few inches to several feet. The cracks are shallow and seldom penetrate below the top layer of reinforcement and are generally numerous.

It is quite common for this type of cracking to occur in hot weather, and also in cooler weather when there is a strong wind, and it can cause consternation for those who do not realize what it is. Unless it is very severe, when it may result in a permanently weakened surface to the slab, it does no

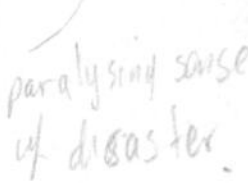

real harm. The cracks should be grouted in with a Portland cement grout well brushed in, and the treated surface should be covered with polythene sheets held down around the edges with planks and blocks. However, prevention is always better than cure and the following are the recommended precautions:

(a) The sub-base and any timber formwork should be well damped prior to placing the concrete.
(b) The aggregates, particularly if they are dry and/or have relatively high absorption, should be sprayed with water.
(c) It may be advisable to use a slightly finer grading of sand, say from zone 2 to zone 3.
(d) An air entraining admixture can often be used with advantage. The air content of the mix should be in the range $4\frac{1}{2}\% \pm 1\frac{1}{2}\%$. This should be strictly controlled by means of frequent checks with an air meter.
(e) The concrete, immediately after compaction and finishing, should be shielded from hot sun and strong wind (see also (g) below).
(f) The placing, compacting and finishing should be proceeded with quickly and without delay between each operation.
(g) Curing should be commenced as soon as possible after finishing is complete. The use of 500-gauge polyethylene sheets, well lapped and held down around the edges with boards and blocks to prevent draughts impinging on the fresh concrete, will give good results. The use of a good quality curing membrane with pigments to reflect light and heat from the sun should also be satisfactory, but some physical screening may be necessary in addition, in extreme conditions.

Plastic Settlement Cracking

This type of cracking is quite different in its origin to plastic shrinkage cracking which has just been described. It can be caused in two different ways. One, by the resistance of the surface of the formwork to the downward settlement (compaction) of the plastic concrete under the action of poker vibrators and the force of gravity. The resistance delays this downward movement, and when it does occur, if the concrete has already stiffened, a crack is very likely to form. This is invariably a surface defect and the crack is wider at the surface and normally does not penetrate deeper than 20–25 mm. The second is more serious, because the cracks penetrate at least to the reinforcement, they may be wider inside the concrete than on the

Fig. 7.1. *Plastic settlement cracking. (Courtesy: H. N. Tomsett.)*

surface, and may be associated with honeycombing beneath reinforcing bars. The cracks are caused by concrete becoming 'hung up' on either the reinforcement or on spacers, and a crack and an associated hollow forms as the concrete stiffens.

Some adjustment in the mix proportions, and the use of a plasticizer or air entraining admixture may solve the problem. However, it may be necessary to use external vibrators on the formwork and to switch these on a short time after the normal compaction has been completed. The time-lag will depend on the rate of stiffening of the concrete, and can vary from $\frac{1}{2}$ h to 2 h.

Figure 7.1 shows this type of cracking near the head of a mushroom-headed column in a reservoir. The author is indebted to H. N. Tomsett of the Cement and Concrete Association for this photograph.

It is advisable for cracks of this type to be repaired by crack injection.

7.5. Thermal Contraction Cracking

General Considerations

During the setting and early hardening process of concrete considerable heat is evolved by the chemical reaction between the water and the cement which results in a rise in temperature of the concrete. The actual rise, the peak temperature, and the time taken to reach the peak and then to cool down, will depend on a large number of factors of which the following are the most important:

(a) The ambient air temperature.
(b) The temperature of the concrete at the time of placing.
(c) The type of formwork used (whether timber, plastic or steel), and the time the formwork is kept in position.
(d) The ratio of the exposed surface area of the concrete, i.e. the area not protected by formwork, to the volume of concrete.
(e) The thickness of the section cast.
(f) The type of cement used and the cement content of the mix.
(g) Whether any provision is made for the thermal insulation of the concrete after the formwork is removed.
(h) The method of curing.

The above are not placed in order of importance because this depends on site conditions.

The detailed behaviour and characteristics of concrete as it hardens are not yet fully known and research on this is being carried out in the UK and other countries. As the temperature of the concrete rises the concrete expands and when it cools down it contracts. The coefficient of thermal expansion (and contraction), depends on a number of factors of which the principal are the type of aggregate and the mix proportions. As the mix proportions of concrete used in liquid retaining structures are likely to vary only within fairly narrow limits, the aggregate type assumes greater importance and this has been mentioned in Chapter 5 and in the sections dealing with the use of limestone aggregates (pp. 103 and 150).

Unless the section (floor, wall or roof) is completely unrestrained (a state of affairs never met in practice), thermal stress will be developed, particularly as the concrete cools down and contracts. The greater the restraint, the higher the thermal contraction stress; these stresses are generally tensile, but in certain parts of a structural member they may be compressive. In the types of liquid retaining structures covered by this

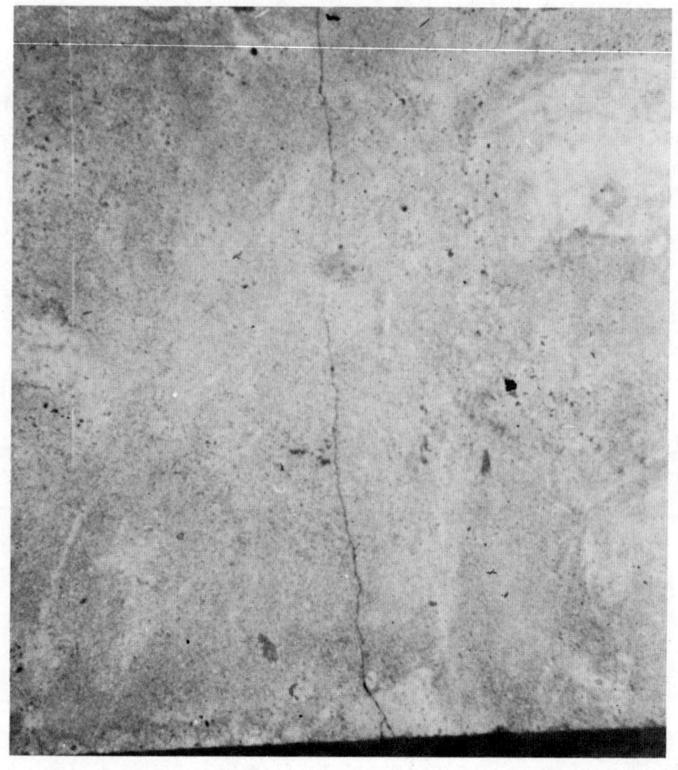

Fig. 7.2. *Thermal contraction crack in wall of sewage tank.*

book, it is the tensile stresses which are important. Figure 7.2 shows a typical themal contraction crack in a wall panel.

The tensile stresses discussed above may, and often do, exceed the strength of the concrete in tension and/or the bond strength of the distribution reinforcement, and cracking will then occur. Measures to reduce the incidence of this thermal cracking may be considered under the following headings:

(a) Measures to limit the rate of temperature rise and the peak temperature reached.
(b) Measures to reduce restraint (resistance) to contraction of the section as it cools and to control cracking.

Each of the above will now be taken in turn and discussed in detail.

Measures to Limit Temperature Rise

Generally, the factors mentioned here are equally applicable to all parts of a liquid retaining structure, i.e. floor, walls and roof. However, in many respects, walls are more seriously affected by temperature rise and subsequent cooling mainly because they are thicker and have formwork on both faces, so that the area of concrete exposed to the external air is relatively small compared with floor and roof slabs.

Timber formwork has a much lower thermal transmittance (U) value than steel; consequently the temperature rise in a wall cast between timber forms will be appreciably higher than a similar wall cast under the same conditions between steel forms. The amount of heat lost through steel formwork may be 20 times that lost through 20-mm-thick plywood or formwork. The actual rise in temperature will depend also on the thickness of the section. In a reservoir wall, say, 450–600 mm thick, the peak temperature in the centre of the wall may reach 65–70 °C, perhaps within 20 h of casting. If timber formwork is used, and this is removed on a cool windy day in summer a steep temperature gradient is formed through the wall and the outer layers of concrete may cool fairly rapidly to 15 °C.

In considering the tensile and bond strength of concrete associated with thermal crack formation, it is advisable to assume a realistic strength for an age of 48 h after casting. The coefficient of thermal expansion (and contraction) of concrete varies between $6{\cdot}3 \times 10^{-6}$ per °C and $11{\cdot}7 \times 10^{-6}$ per °C which gives an average of $9{\cdot}0 \times 10^{-6}$ per °C. Concrete made with a limestone aggregate is likely to have a coefficient of thermal expansion of about $7{\cdot}5 \times 10^{-6}$ per °C.

In the types of structure considered in this book only Portland cements would be used. The rate and amount of evolution of heat and hydration will vary with different types of Portland cement.

Low heat Portland cement (BS 1370) would give the most advantageous results; the British Standard requires that the heat of hydration developed over 7 days and 28 days must not exceed 60 cal/g and 70 cal/g respectively. The same restrictions on heat of hydration apply to low heat Portland blastfurnace cement (BS 4246). However, it is unlikely that either of these cements would be used for liquid retaining structures due to difficulties in obtaining regular supplies in relatively small quantities.

Some sulphate resisting Portland cements (BS 4027) will generally develop less heat in a given period than ordinary Portland, but the difference is not great, and the cement manufacturer should always be consulted before a decision is made to use sulphate resisting instead of ordinary Portland cement for its low heat properties alone.

Portland blastfurnace cement (BS 146) will generally develop less heat in a given period than either ordinary Portland or sulphate resisting Portland cement: its availability is, for practical purposes, limited to the area where it is made in Scotland, unless very large quantities are required.

Apart from the type of cement, the quantity of cement per unit volume of concrete is also important; the higher the cement content the higher the peak temperature and the more rapid the rise in temperature.

Measures to Reduce Restraint and to Control Cracking

Generally speaking, concrete floor slabs and wall panels would not develop thermal contraction cracks if they could be constructed so as to be completely unrestrained. It is the restraint which gives rise to the stress. A bay in a floor slab is restrained on the undersurface, and around the perimeter if it is bonded to adjacent bays. A wall panel is restrained at the base where it is bonded to the kicker, and also to adjacent panels if the distribution steel is continuous from one panel to another, and/or efforts are made to form a monolithic joint to an adjacent previously cast panel.

Roof slabs are restrained by supporting beams which in turn are usually restrained by supporting columns or walls unless a sliding joint is provided.

Having recognized that there is restraint, the problem arises as to how this can be reduced, provided always that structural stability, durability and watertightness are not impaired.

FLOOR SLABS UNIFORMLY SUPPORTED ON THE GROUND

The friction developed on the under-surface of the slab can be reduced by casting the slab on a double layer of polyethylene sheets (500 gauge) which themselves are laid on the oversite concrete. To reduce restraint, the perimeter bays, which in reservoirs support the walls, are not tied to the adjacent floor bays. This can be achieved by forming a plain butt joint longitudinally, and stopping off the reinforcement. In structures such as swimming pools, where the walls are much less massive, it is usual to tie the perimeter bays to the floor slab bays. If this is not done, there is a tendency for the walls to move outwards. If the floor bays are rectangular, the transverse joints can be either formed by stop ends or crack inducers. In both cases, it is recommended that the reinforcement is not carried across the joint, thus forming a contraction or stress-relief joint. The surface of the concrete at contraction joints formed by stop ends should be left as it comes from the form. If for any reason the surface is rough, then a coat of grout or bituminous emulsion which should be allowed to dry will effectively break bond with concrete of the next bay when it is cast.

The object of a crack inducer is to form a stress-relief joint in such a way that concreting is not held up, and this enables long lengths of floor to be concreted in one continuous operation. Figure 7.3 shows a contraction joint in a floor slab.

The next problem to be settled is the optimum bay size as this will determine the location of the joints in both directions. The bay size and shape (whether square or rectangular) is governed by a number of factors

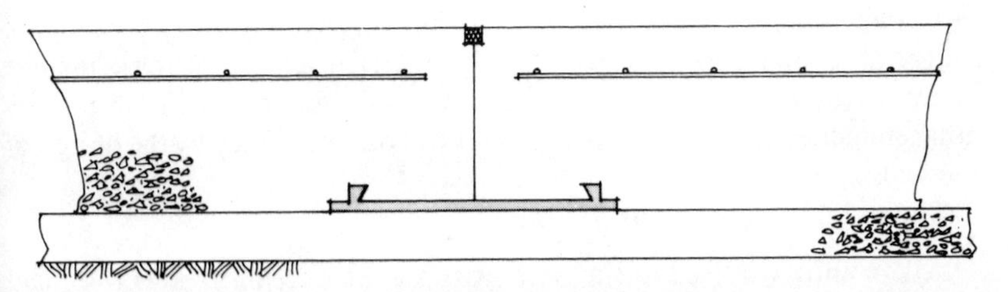

Fig. 7.3. *Contraction joint in floor slab.*

including the position of any columns and the method of compacting the concrete. If the concrete is to be compacted by tampers working off the side forms, then the bay width is limited to about 4·0 m. Subject to column spacing (if there are columns) the length of the bay is largely dependent on the weight of reinforcement provided in the top of the slab, as it is this which controls cracking.

The reinforcement is best provided in the form of a fabric to BS 4483. Rectangular bays should have rectangular fabric and square bays square fabric. It is normally placed 40–50 mm from the top of the slab.

In the majority of liquid retaining structures, the perimeter bays of the floor are in fact the bases for the walls, and the reinforcement from the walls is anchored into the base when the walls are designed as vertical cantilevers. As previously stated, with massive wall sections the joints along the perimeters of the bays should be detailed to allow the bays to contract, and therefore the reinforcement is not carried across the joints. Exceptions to this are those cases where the walls are less massive, and/or the floor is subject to uplift from ground water pressure, and then either the reinforcement is carried across the joints or tie bars are used. This increases the risk of contraction cracking within the bays, but this risk can be offset by increasing the weight of reinforcement.

An example of the design of a ground floor slab is given in Part 1 of this book.

In the conclusion of this section it is relevant to remark that the authors see no advantage from the point of view of controlling cracking, in casting the concrete in alternate bays nor in a checker board pattern. If a contractor wishes to do so for reasons best known to himself there is no objection. The methods described above should provide adequate crack control and at the same time allow concreting to proceed expeditiously.

WALLS

Wall panels are more complicated to deal with than floor slabs which are uniformly supported on the ground. The panels, when they are cast, may be restrained at the base and they may also be restrained at one or both ends by adjacent panels, piers or buttresses. This will be determined by the design of the wall.

The usual designs are as follows:

(a) Cantilever fixed at the base, with the main reinforcement vertical and the distribution steel horizontal. The structure can be square, rectangular or circular on plan.

(b) (i) Structures which are circular on plan with the main reinforcement horizontal in the form of hoop steel.

(ii) Structures circular on plan with part of the wall having the main steel vertical and the upper part having the main reinforcement horizontal in the form of hoop steel.

(c) Structures which are square, rectangular or circular on plan with the wall panels designed as slabs spanning horizontally between piers or buttresses, so that the main reinforcement is horizontal.

Examples of design of some of the above types of walls are given in Part 1 of this book.

(a) Where the walls are designed as cantilevers fixed at the base with the main reinforcement vertical, the panels are usually restrained along their bottom edge where they are bonded to the kicker; in addition they will also be restrained at one or both ends if they are bonded to adjacent panels.

Restraint at the bottom edge can be reduced by casting the wall panel and the base in one continuous operation. While this procedure is unusual it has been done successfully in a number of reservoirs. Informaton obtained from the designers showed that by including this requirement into the contract at tender stage as an alternative to casting the base and wall panels separately, no increase in price was recorded in the successful tender. It is obvious that the formwork has to be designed differently for the two methods and special arrangements made for erection and removal. A further point of interest is that it was found that a top shutter was not required to the upper sloping surface of the base and there was no 'spewing' of the concrete when the wall panel was cast immediately following the casting of the base. The workability of the concrete was carefully controlled in the range 25–50 mm; and in the cases known to the authors the concrete was site mixed.

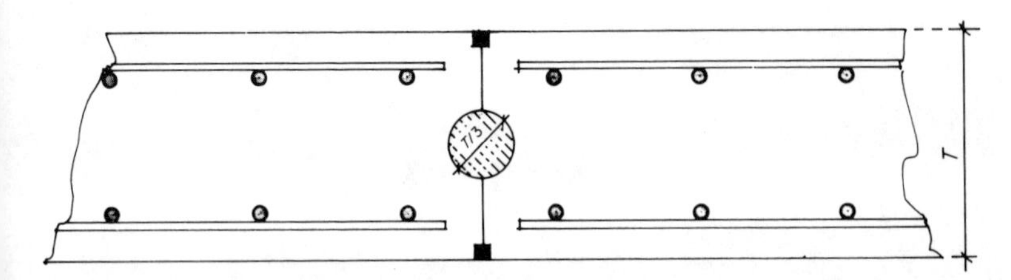

Fig. 7.4. *Stress-relief joint in wall using void former.*

Similar problems of restraint will arise if the wall panels are cast in two vertical lifts and this is an important reason why full-height casting is advantageous.

In the case of the vertical joints between adjacent panels the authors recommend that in most cases these should be detailed as contraction or stress relief joints and not as construction or monolithic joints unless there is a sound structural reason for doing so. If a full contraction joint is provided this will necessitate the fixing of a stop end, and should the designer provide for continuity of steel across the joint the stop end would have to be perforated, which may result in grout loss and honeycombing.

By using crack inducers at appropriate centres, long lengths of wall can be concreted in one continuous operation. Figures 7.4 and 7.5 show stress-relief joints with alternative types of crack inducer. The method using the special type of water bar is generally to be preferred.

The authors feel that the distribution steel can be completely discontinuous at these joints (as shown in the figures). It should be remembered that the percentage of distribution steel used to control thermal contraction cracking is usually quite small unless the design includes for monolithic construction. Therefore this small percentage of steel has little value as shear reinforcement.

Contraction joints which are formed with a stop end should be plain butt joints and no attempt should be made to bond the new concrete to the old.

With stress-relief joints the object is to form a plane of weakness through the wall for the full height at that point. As the concrete cools down and contracts, a crack should form in the groove provided, thus relieving thermal stress in the panel as a whole, and cracks should not occur at intermediate positions within the panel. The groove can be sealed in the usual way.

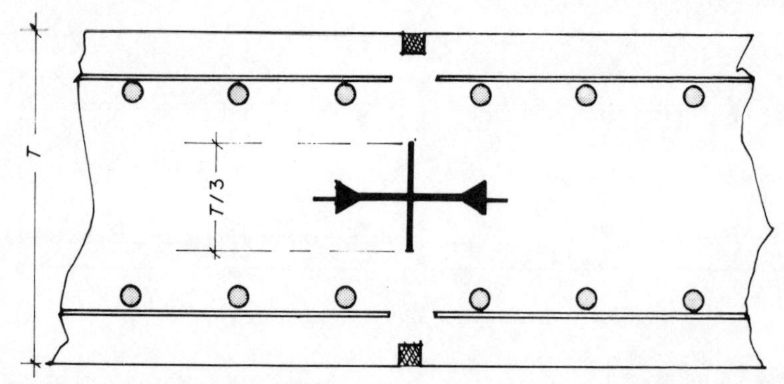

Fig. 7.5. *Stress-relief joint in wall using proprietary water bar. (Courtesy: Expandite Ltd, London.)*

Having taken appropriate measures to reduce restraint when the concrete in the wall cools down and contracts, as discussed on p. 178, consideration should be given to controlling the cracks so that their width does not exceed an acceptable figure.

Normally, cracks arising from thermal contraction are not important structurally, but nevertheless they can result in leakage. The maximum acceptable crack width is a matter on which there is a wide divergence of views, both in this country and abroad.

It should be noted that these cracks are 'parallel' walled cracks, and extend right through the wall. The authors consider that in this context a crack means a crack visible to the naked eye (which includes a person wearing glasses). A crack about 0·125 mm (5 thou in) wide is usually visible in good light; a crack 0·05 mm (2 thou in) wide can be seen in good light but it takes a considerable effort and some experience to find it initially. Once it is seen and marked, then persons subsequently examining the wall can usually see it without undue difficulty. In addition to these remarks which are based on the authors' personal experience, there is the problem of actually measuring the width of such fine cracks. Even cracks 0·25 mm (10 thou in) wide are difficult to measure as the crack follows a very irregular path and meanders around pieces of fine and coarse aggregate. The width of cracks can be measured in three basic ways:

(a) By the use of a crack width microscope.
(b) By the use of a set of feeler gauges. These are used in the same way as when measuring the electrode gap in sparking plugs.
(c) By the use of a special plastic plate on which are ruled lines of standard thickness (i.e. from 0·1000 mm to 2·00 mm). These are 'matched' with the crack (Colebrand Ltd, patent applied for).

While the microscope is undoubtedly the most refined method, it is not really suitable for most site conditions and the authors' experience is that method (c) is probably adequate for the majority of cases met in practice.

It is unrealistic to argue about exact crack width. The majority of liquid retaining structures have walls which are designed as cantilevers fixed at the base and the need to provide for thermal contraction stresses has already been discussed. To provide for cracking at predetermined positions requires a combination of experience in design and in construction, so that when the structure is built, there will be little, or if one is lucky no, intermediate cracking.

(b) (i) In the walls of circular tanks the main steel is horizontal, i.e. hoop steel. The provision of a sliding joint between the wall and the floor will substantially reduce restraint at that point, and will allow the wall to move inwards, i.e. reduce in length, as the concrete cools down and contracts.

In theory it should be possible to concrete the whole of the wall of such a tank in one continuous operation, and the authors have read of cases where this has been done successfully, particularly when slip form construction techniques have been used. The hoop steel has to be continuous and so if the wall is cast in panels the horizontal steel will cross the joints. A conservative approach to this problem is to cast the walls for their full height in panel lengths of about 6–7 m. This can probably be safely increased to 10 m if there is a well designed and constructed sliding joint at the base of the wall. If thermal contraction cracking does occur, then the cracks can be sealed by crack injection. The use of PTFE for such sliding joints is recommended as this material has a very low coefficient of friction.

Water bars at the base of the wall are usually rubber as this is considered more durable under frequent movement.

(b) (ii) With circular tanks which have the walls fixed to the base, the stress distribution is rather complicated and the lower part of the wall has the main reinforcement vertical, while from a certain height above the base,

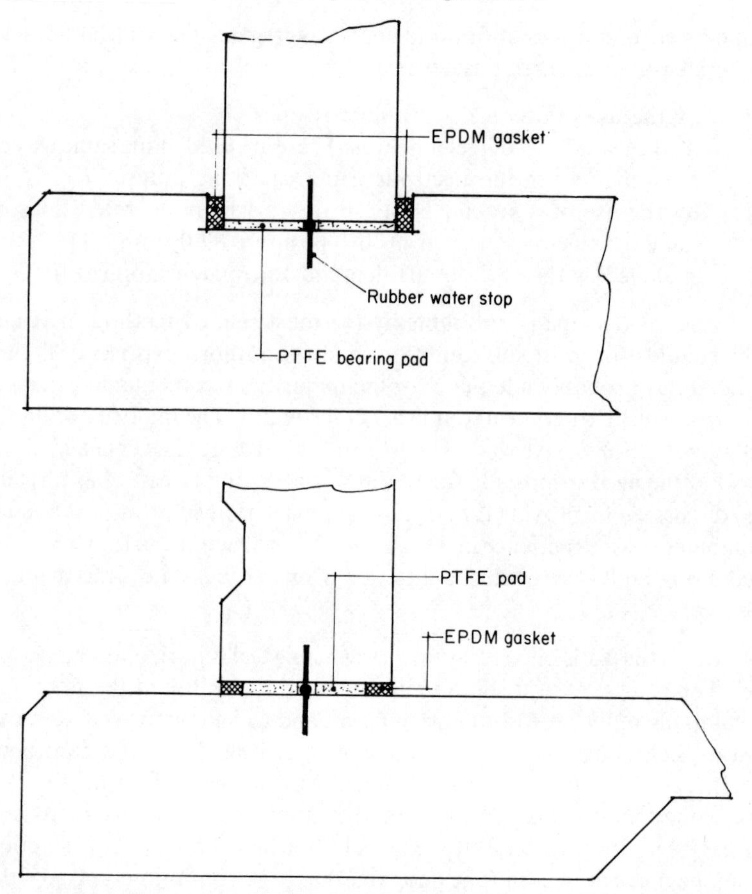

Fig. 7.6. *Sliding joint between wall and floor of circular tank.*

ring tension becomes predominant and the main reinforcement is in the form of hoop steel, which is usually quite substantial in area.

As in the case of circular tanks with a sliding joint at the base (see Fig. 7.6), the hoop steel cannot be reduced or curtailed when it crosses vertical joints between one panel and the next. In these circumstances the authors recommend that the walls should, for the purpose of deciding on the panel length, be considered as cantilevers fixed at the base, and the panel length be determined on this basis. This is admittedly conservative, because in the upper part of the wall there is a substantial area of hoop steel.

The casting of the base and the wall panel in one continuous operation will undoubtedly help to reduce the possibility of cracking, otherwise the new concrete in the wall panel is being bonded to a relatively cool, contracted base which will induce considerable restraint on the thermal contraction of the newly cast panel.

Any vertical cracks which form due to thermal contraction can be repaired by crack injection.

(c) Where the walls are designed to span horizontally between piers or buttresses, it is considered that a monolithic joint between the panel and the floor slab is not really necessary, but will do no harm. This is because the restraint induced by such a monolithic joint will be small compared with the restraint imposed by the piers or buttresses at each end of the wall panel. It is, however, reasonable to assume that a monolithic joint in this position is easier to make watertight than an unbonded joint; the concrete can be well compacted against the prepared surface of the kicker and a steel water bar used from which positive bond with the concrete will be obtained.

This type of construction is rather unusual except in elevated tanks.

Roof Slabs

Difficulties can and do arise with thermal contraction in roof slabs unless they are post-tensioned which is unusual for the roof of a liquid retaining structure. Assuming that the design is adequate to cater for the external loads, the trouble arises from the contraction of the concrete as it cools down and later drying shrinkage may cause the cracks to widen.

As has been pointed out with floors and walls, if the roof slab were supported on a frictionless bearing, the restraint would for all practical purposes be zero and these secondary stresses would be entirely insignificant. In practice, restraint on the contraction of the roof slab is imposed by supporting beams, columns and walls.

In recent years problems associated with this type of cracking have increased, due mainly to the pressure from contractors to cast these suspended slabs in larger and larger areas in one continuous operation. It is now not unusual to cast 300 m^3 of high quality concrete in one 'pour'.

Of the two factors, thermal contraction and drying shrinkage, the former exercises the predominant influence.

While alternate bay and chequer board construction has nothing to recommend it for ground floor slabs, this does not necessarily apply to thick suspended slabs where the heat build-up may be considerable. In these cases there may well be an advantage in delaying the casting of adjacent bays until the concrete has cooled and a considerable percentage of the initial thermal contraction has taken place. Generally, a period of 72 h is adequate for this.

In the case of reservoirs holding potable water, it is usual for the roof slab to be designed and constructed so as to be watertight. Under certain conditions, e.g. timber formwork, large spans, thick slab, and high ambient air temperature, there is likely to be an appreciable rise in temperature of the concrete. With suspended slabs it is not possible at the present time to predict with any degree of precision, the magnitude of these thermal contraction stresses. The factors affecting the temperature rise have already been discussed earlier in this chapter. The problem is to try to assess the effect of the restraint imposed by the supports in each particular case. For example, if the slab is supported on walls with a sliding joint, restraint will be much reduced compared with a fixed joint in this position. Similarly, massive beams on massive, heavily reinforced columns will introduce considerably more restraint than lighter beams on slim, flexible columns. If it is required to cast a large area of concrete in one operation the following precautions should be taken:

(a) All the measures which are relevant to reduce the rise in temperature of the concrete and which have been previously discussed should be taken.
(b) The whole concreting operation should be carefully pre-planned. If the concrete is ready-mixed, the complete understanding and co-operation of the supplier should be ensured by prior consultation, well in advance of the concreting operation.
(c) An adequate amount of experienced supervision should be provided on site.
(d) Within the area of one bay, i.e. one casting unit, there should be no part of the slab without adequate top reinforcement in the form of a fabric (BS 4483). The fabric is placed 40–50 mm from the top of the

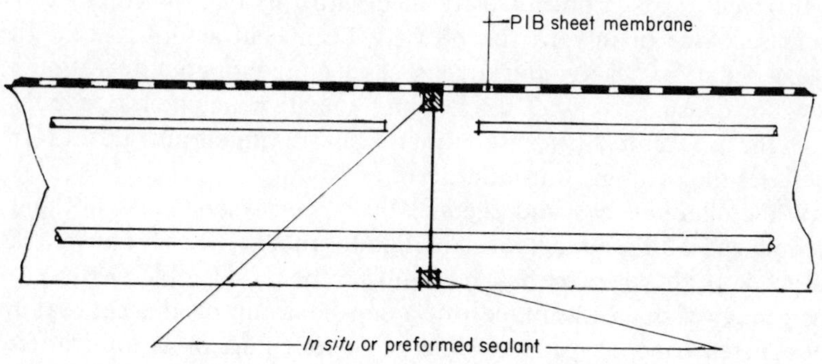

Fig. 7.7. *Contraction joint in roof slab.*

slab and is additional to the reinforcement required for normal structural design purposes.

(e) Stress-relief joints should be provided at selected locations. These are vertical planes of weakness purposely formed in the slab in positions where their presence will not reduce the resistance of the slab to external loads, under the worst conditions of loading. Their location should be determined in each case as a separate design problem. A suggested detail for such a joint is shown in Fig. 7.7.

A full movement joint in a roof slab is shown in Fig. 7.10 (p. 190).

7.6. *Joints*

General Considerations

All engineers know that joints of some description are needed in liquid retaining structures, but the reasons behind the correct selection of the type of joint, its location, and its detailing are often not clearly understood. There are three basic reasons for providing joints:

(a) It is not practical to cast the whole structure in one continuous operation.
(b) Generally, the larger the volume of concrete cast the higher are the stresses produced in the concrete by thermal contraction in the early life of the concrete, and later by drying shrinkage. These secondary stresses can be controlled by the judicious use of selected types of joints.
(c) Certain types of joints, such as pin joints, sliding joints and full movement joints perform an essential role in the basic structural design of the structure.

An essential requirement for all joints is that they must be watertight. Detailed information on joints sealants, inert fillers and water bars is given in Chapter 5. Apart from full movement joints, which usually traverse the structure from one side to the other and extend from the floor to the roof, the usual position of joints are

(a) between the floor and walls, and floor and column bases;
(b) between column bases and columns;
(c) between walls and roof and columns and roof;
(d) between bays in floor and roof slabs;
(e) between panels in walls.

Joints are dealt with in section 5 of the Code (BS 5337). The reader is referred to Chapter 11 for comments on the Code.

Having agreed that joints will have to be provided in the locations indicated above, the next problem is to decide on the type of joint and this decision will have important effects on the design of the structure, on its behaviour during construction, and throughout its life when it is subject to variations in load, foundation movement, temperature and moisture changes, etc.

There is a detailed discussion on the causes, effects and suggested solutions to problems of thermal contraction cracking in the previous section of this chapter and so this will not be repeated here.

Joints may be conveniently divided into the following types:

(a) Full movement joints; these are often loosely termed expansion joints.
(b) Partial movement joints; these are also referred to as contraction, partial contraction, and stress-relief joints.
(c) Sliding joints.
(d) Monolithic joints; these include construction and daywork joints.
(e) Pin joints.

Full Movement Joints

These joints should accommodate expansion and contraction on both sides of the joint. The basic requirements are that there should be no structural continuity across the joint, the joint should be able to open and close with the minimum of restraint, and it must remain watertight under all anticipated conditions of movement throughout its life. Generally such joints are continuous right across a structure and extend from floor to roof. This is to ensure complete separation between the two adjacent parts of the structure. Full movement joints are used extensively in liquid retaining structures built in areas subject to mining subsidence. In such cases vertical and horizontal movements may be considerable, and the structure itself may show a vertical displacement from one side to the other of more than 500 mm (20 in).

Figures 7.8 and 7.9 show details of full movement joints in the floor and wall of a reinforced concrete structure. Important points to note from these diagrams are given below.

FLOORS

A water bar is provided but is preferably placed below the main floor slab on

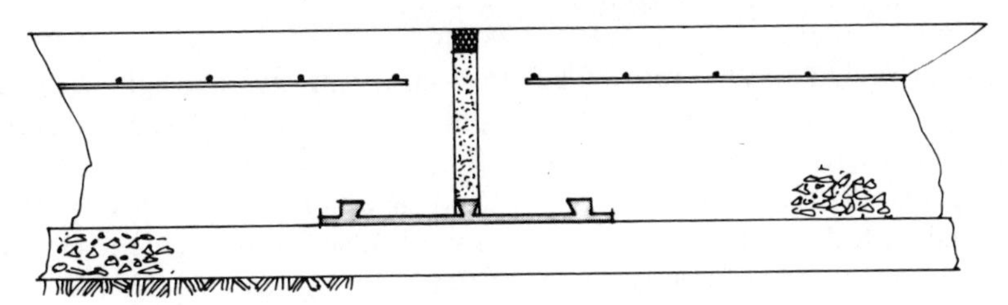

Fig. 7.8. *Full movement joint in floor slab.*

the oversite concrete, rather than in the middle depth of the slab. The reason for this is that it is difficult to ensure proper compaction of the concrete around a horizontal centrally placed water bar.

An exception to this externally located position is in floors of structures subject to considerable ground movement (e.g. mining subsidence areas); in such cases it is usual to place the bar centrally in the depth of the slab and take special care with the compaction of the concrete.

Walls

It is usual for the water bar to be centrally located. In vertical joints this requires the stop end to be split to allow the water bar to pass through; care must therefore be taken to ensure a grout-tight joint. It is important for the wall to be sufficiently thick to allow proper compaction of the concrete around the water bar. In large reservoirs and tanks this presents no difficulty, but in small structures designers sometimes calculate a minimum thickness for structural stability and forget the practical aspects of obtaining compaction of the concrete. In the opinion of the authors, wall thickness, wherever practical, should be at least 250 mm, although it is recognized that thinner walls might be used for economic reasons, particularly in small projects.

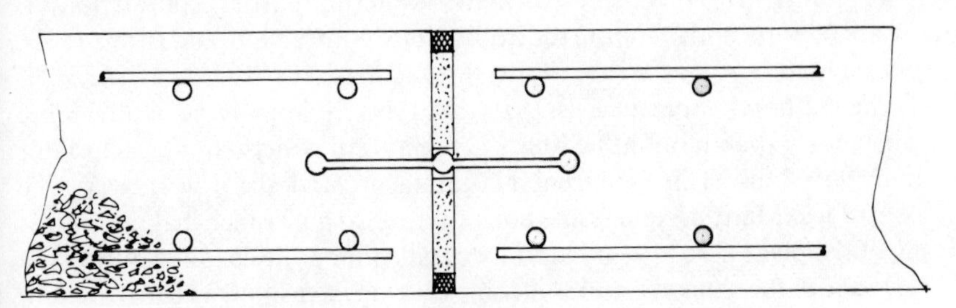

Fig. 7.9. *Full movement joint in wall.*

ROOFS

In some ways it may be more difficult to ensure complete watertightness in a full movement joint in a roof slab than in a floor or wall. It is not usual to provide the necessary special detailing for the provision of an external type water bar on the underside of a roof slab so the water bar has to be centrally placed with the resulting difficulty in the compaction of the concrete. A detail for this type of joint is shown in Fig. 7.10. It is emphasized that the

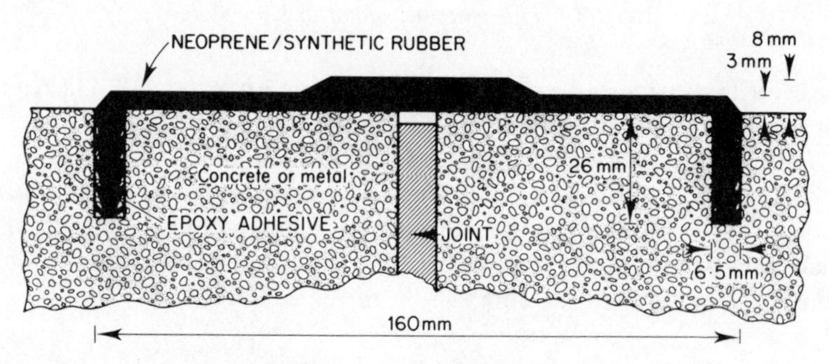

Fig. 7.10. *Channel section gasket used to seal full movement joint in roof slab. (Courtesy: Colebrand Ltd, London.)*

designing engineer must use his professional judgement in deciding on the details of these joints.

If the roof is provided with a waterproof membrane, then the sealing of joints in the slab is facilitated and except in the case of full movement joints, can probably be dispensed with.

Partial Movement Joints

GENERAL

This category of joint includes contraction, partial contraction, and stress-relief joints. It is considered that the term 'partial contraction' is unsatisfactory, and therefore the expressions 'contraction' and 'stress-relief' will be used here.

The authors' experience is that this type of joint is generally more satisfactory than monolithic joints between wall panels, and bays in floor and roof slabs. This is a general statement, and there will always be exceptions. Many designers and constructors do not realize that movement may take place across their carefully detailed monolithic joints during the early age of the concrete, and sometimes during later stages of construction, during filling and emptying for testing, and during the normal operation of the structure. For example, in reservoirs, it is quite common for the walls to be 60 m (200 ft) or more in length between full movement joints. With panel lengths of 5·0 m (16 ft), it is unrealistic to assume that considerable stress will not develop across these joints after the walls have been cast and during filling and emptying.

In the case of more specialized structures, such as swimming pools, and tanks holding heated liquids, there are likely to be appreciable changes in stress and strain during the normal operation of the structure. This fact was demonstrated in papers by Southern and Hodgson at the Symposium on Swimming Pools sponsored by the Building Research Establishment and held at East Kilbride on 6 June 1973.

If the designer feels that structural continuity is necessary across joints between wall panels and floor and roof bays, then he should give consideration to the provision of a water bar and/or sealing groove, as a safeguard against possible opening of the joint. Figures 7.3, 7.4, 7.5 and 7.7 show details of partial movement joints in a floor slab, a wall, and a roof slab. The following notes amplify the diagrams.

FLOORS

The use of a stress-relief joint which incorporates a crack inducer enables long lengths of slab to be cast in one continuous operation. This can be particularly useful on large projects where the concrete is pumped. An external type water bar is provided and types are now on the market which incorporate a crack inducing projection. The sealing groove should be formed in the plastic concrete as the work proceeds as this is a safer procedure than cutting the groove after the concrete has hardened. It is always difficult to decide on the optimum time for sawing the joints as this can vary from day to day, depending largely on ambient air temperature. If sawing is carried out too early after concreting, ravelling of the joint can result. Some aggregates, particularly flint gravels, are very difficult to saw. If the cutting is delayed, the slab may have already cracked and such cracks are much more difficult to seal than a groove.

Stop ends are used to define the limits of each casting sequence. When the stop end is removed, the surface of the concrete at the joint should be simply rubbed down; a thin coat of paint or bituminous emulsion can be applied to reduce bond between the old and new concrete.

WALLS

The two ways of forming partial movement joints in walls are the same as in floors, namely by the use of stop ends or by crack inducers. Both can be equally effective, but if a centrally placed water bar is used, the stop end has to be split to allow the water bar to pass through and this can result in grout leakage and consequent honeycombing in the concrete adjacent to the water bar. Full compaction of the concrete around the water bar is essential for watertightness.

A stress-relief joint can be formed effectively with a centrally placed water bar as shown in Fig. 7.5. However this type of joint is often formed without a water bar by means of an inflatable centrally located void former, and this is shown in Fig. 7.4.

Care must be taken to fix the inflatable former in position so that it is not displaced during the placing and compaction of the concrete. Also, the plane of weakness created in the wall must be at right angles to the face and pass through the centre line of the sealing groove(s) and the inflatable former. Unless all horizontal steel is stopped off 40 mm ($1\frac{1}{2}$ in) on each side of the joint, then not less than 50 % of this steel should be accurately cut on the centre line of the joint. After concreting, the inflatable former is removed and the hole must be immediately and carefully plugged. Provision must be made for the cleaning out of the void before it is finally sealed, and for this purpose a hole about 25–40 mm (1–$1\frac{1}{2}$ in) diameter is provided, sloping from the void to the outside or inside face of the wall.

The final sealing of the void should be carried out as late in the construction process as possible. Materials for sealing can be flexible or rigid, depending on whether movement across the joint is anticipated during the normal operation of the structure.

The use of the special water bar is generally to be preferred.

The flexible compounds are usually hot-applied thermoplastics, such as rubber-bitumen, but cold-applied elastomers such as polysulphides and polyurethanes can also be used, but as the volume required would be considerable, the cost is likely to be prohibitive. For more information on these materials, reference should be made to Chapter 5.

The rigid materials would be low-shrink grout mortar, or cement mortar containing a rubber latex emulsion based on styrene butadiene.

The sealing groove(s) should be sealed in the usual way. If the work is carried out carefully, a watertight joint should result. Unfortunately, the flexible sealants are relatively short lived compared with the concrete and have to be renewed periodically. From what has been said, it can be seen that the correct use of a specially designed water bar is a much simpler way of forming a stress-relief joint in a wall than the use of a void former.

If it can be reasonably assumed that once the structure is complete and in operation further movement at these stress-relief joints will not take place, then in the authors' opinions there is no valid reason why the joints should not be 'locked' by sealing them with an epoxide mortar. The latter is likely to have a much longer life than a flexible sealant.

Roofs

Some of the practical problems of stress-relief joints in roof slabs have been discussed on previous pages as well as the waterproofing of full movement joints.

Figure 7.10 shows a suggested detail for a full movement joint, and resourceful designers will no doubt produce others to suit particular circumstances. The same accuracy and care is required as in floor and walls, otherwise the crack may not form at the selected location (which is within the sealing groove) and thus make it more difficult to seal effectively. Roofs can impose certain problems because they are suspended slabs and so are different in design to slabs which are uniformly supported on the ground. This is discussed in Chapter 8.

Sliding Joints

General

The two places where a sliding joint is likely to be used are

(a) between the walls and floor of circular reinforced and prestressed post-tensioned concrete liquid retaining structures;
(b) between the roof slab and the walls and/or beams of reinforced and prestressed post-tensioned concrete liquid retaining structures.

The fundamental requirements for this type of joint in the structures mentioned above are

(a) the frictional resistance between the contact faces of the joint should be as low as possible;
(b) the joint, depending on its position, may be required to remain watertight under all anticipated conditions of service.

Sliding Joints Between Floor and Wall

As stated above, this type of joint is often used on circular structures, both reinforced and prestressed. In the case of reinforced structures, it is only used when the walls are designed as cylinders with the main reinforcement horizontal (hoop steel). A sliding joint in this position is advantageous from the point of view of reducing restraint at the base of the wall; this has been mentioned in Section 7.4 dealing with thermal contraction.

Figures 7.6 and 7.11 show sliding joints.

The use of a rubber water bar rather than PVC is recommended. The area of the concrete floor which supports the wall should be finished as smooth as possible. This can be achieved by trowelling-in a thin layer of cement mortar before the concrete has hardened, and finishing this with a steel float. Then, a few days before the formwork for the wall is erected, the surface is given two coats of bituminous enamel. The enamel should be hard

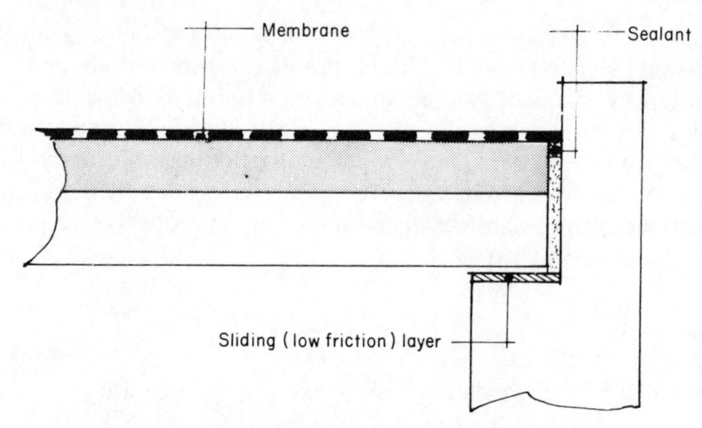

Fig. 7.11. *Sliding joint between roof slab and wall.*

and smooth when the concrete is cast on it. A more expensive method is to use a special layer of low friction material such as polytetrafluoroethylene (PTFE).

It must be admitted that with this type of joint it is difficult to ensure a grout-tight joint at the base of the wall when the concrete is cast as there is no kicker against which the formwork can be tightened.

In the case of structures where the walls are prestressed circumferentially, as in the Preload system, a sliding joint between the floor and wall is essential. The Preload joint utilizes a foam rubber pad to form the sliding layer. Further information on prestressed tanks is given in Chapter 9.

Sliding Joint Between Roof Slab and Supports

There is a divergence of views as to the advantages and disadvantages of a sliding joint between a roof slab and its supports. It certainly reduces restraint in the roof slab and the advantages of this in reducing the possibility of thermal contraction cracking has been mentioned in Section 7.4. It also simplifies the design of the wall. On the other hand it may be necessary to provide a watertight joint between the roof slab and the wall. At internal positions, such as a slab simply supported on beams, watertightness may not be an important factor, depending on the detailed design.

Figure 7.11 shows a sliding joint between roof and wall.

Irrespective of the position of the joint, the surface of the supporting concrete must be finished as smooth as possible and the methods previously described for the floor to wall joint can be used as well as a double layer of 1000-gauge polythene sheeting laid on a smooth mortar bed.

Monolithic Joints

Unfortunately, monolithic joints are often provided in positions where a contraction or stress-relief joint would be more suitable. The result of this is frequently the occurrence of cracks, particularly in walls.

This type of joint should only be used where the designer requires structural continuity, and in theory the joint is assumed to be as monolithic as the structural element (wall, floor, roof) itself.

There are, however, occasions during construction where concreting operations have to be suspended between predetermined contraction, stress-relief or movement joint positions. In such cases, a monolithic joint, also called a construction or daywork joint, is the logical solution.

In the case of vertical joints in wall panels, and joints between bays in floor and roof slabs, it is best to use a temporary stop end. For horizontal joints in walls, the concrete should be brought to a level surface. Sloping surfaces to the joint should be avoided.

The joint between the kicker and the wall panel, which is common to most liquid retaining structures, is a monolithic joint. The monolithic nature of the joint depends on obtaining the best possible bond between the old concrete and the new concrete when it is compacted against it. All reinforcement should be carried across the joint.

To obtain good bond the following work should be carried out:

(a) The surface of the 'old' concrete for the full thickness of the slab or wall should be prepared so as to obtain a mechanical key for the new concrete. This preparation can consist of bush hammering the concrete after it has hardened. An alternative is to use a retarder on the stop end and strip this as soon as possible and then wire brush or grit blast the surface to expose the coarse aggregate. For the surface of a kicker, or a horizontal construction joint in a wall panel, the aggregate can be exposed by the use of a water spray and brush before the concrete has hardened. If this is done properly, it often gives better results and is quicker than brush hammering the mature concrete, particularly if there is a congestion of reinforcement at the joint.

(b) The new concrete should be carefully compacted against the old.

It is the opinion of the authors that there is no need to use a layer of grout or mortar. In fact, if there is delay between the application of the grout/mortar and the placing of the new concrete, a de-bonding layer can be formed, resulting in considerable leakage when the water test is applied.

When concreting a kicker joint or a horizontal construction joint in a wall panel, some engineers specify that the first 100–150 mm depth of concrete should be a richer mix than for the remainder of the 'pour'. This generally means in practice an increase in cement content of about 50 kg per cubic metre of concrete.

In the previous section dealing with partial movement joints, mention was made of the recorded movement which takes place across joints in swimming pools, due largely to temperature changes when the pool is filled and emptied. There is no reason to doubt that the same state of affairs exists with other structures holding heated liquids. Therefore when monolithic joints are provided for reasons of structural continuity, it is wise to consider the inclusion of a water bar and sealing groove; this will help ensure watertightness should the joint open.

When a flexible type water bar is used in a kicker joint or other horizontal joint in a wall, this should be of the type which can be securely laced to the reinforcement so as to prevent displacement during the placing of the concrete.

Pin Joints

Pin joints are sometimes referred to as hinged joints and when used in liquid retaining structures they are sometimes located at the head of columns and rather more often at the base and intermediate positions. Such a joint is a design concept and is used in the design calculations for the columns.

A hinge has been defined by Charles Reynolds in his *Reinforced Concrete Designers Handbook* as

> an element that can transmit thrust and a shearing force but permits rotation without restraint

Figure 7.12 shows a typical pin joint at the base of a column in a reservoir.

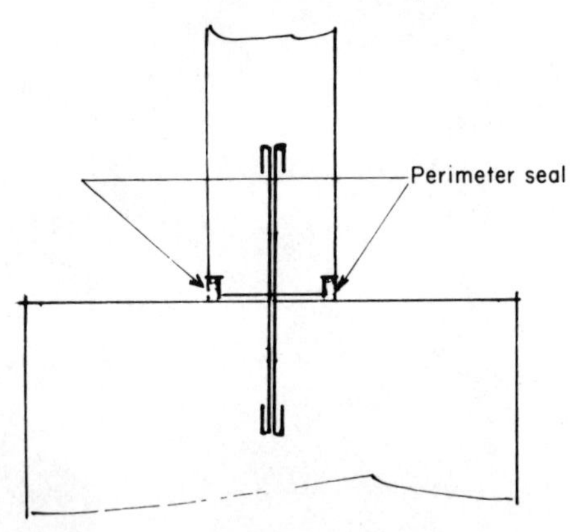

Fig. 7.12. *Pin joint at base of column.*

7.7. *The Use of High Velocity Water Jets and Oxy-acetylene Flame for Preparing Concrete Surfaces and for Cutting Concrete*

The use of high velocity (high pressure) water jets for a variety of mechanical engineering purposes, particularly in the marine field, has been in existence for the last 15–20 years. However, apart from the cleaning of structural steelwork, it is only in the past few years that this technique has been used in the construction industry.

Work by the Construction Industry Research and Information Association (CIRIA) showed that high velocity water jets are as effective in cleaning steelwork as conventional methods of shot and grit blasting. It appears that the high velocity water removes practically all traces of aggressive chemicals from the surface of the steel in addition to the complete removal of rust and mill-scale.

It is a logical step to extend the field of application to concrete. The technique can be used for removing the surface and exposing the coarse aggregate, removing deteriorated or defective concrete and for cutting concrete.

The high velocity jets work without much noise (except from the machinery) and there is no vibration and no dust. For work on concrete the pressure required to produce the necessary nozzle velocity is about 400 atm or 6000 lbf/in^2. The quantity of water used is quite small, about 55 litre/min (12 gal/min) and of this, about 30 % is dissipated as spray. The surface of the

Fig. 7.13. *Use of high velocity water jets for preparing and cutting concrete.*

Fig. 7.13.—*contd.*

concrete is left clean and damp which is ideal for achieving good bond with subsequent layers of concrete or mortar. Figure 7.13 shows the equipment in use and an example of how a reinforced concrete wall about 250 mm thick was cut in order to effect a junction with a new wall. The wall was 6·0 m high and the cut was tapered from 300 mm wide on the outer face to 50 mm at the rear; the work was completed on $7\frac{1}{2}$ h.

A method new to the UK, but one used extensively in Germany and elsewhere on the Continent, for cleaning and preparing the surface of concrete to receive subsequent layers or coatings which have to develop good bond with the concrete is the use of oxy-acetylene flame. This technique has the advantage that there is no noise and vibration and very little dust. The flame spalls or textures the surface rapidly; the speed of travel usually adopted is between 1·2 and 2·0 m/min. The depth of exposure of the coarse aggregate is largely determined by the speed of travel of the flame, although the properties of the concrete including the moisture content of the surface layers are also important.

The flame temperature is about 3000 °C, but the temperature reached by the concrete below a depth of about 2 mm does not exceed 150 °C with normal rates of travel. While the draught created by the passage of the flame will remove some of the surface scale, final brushing with a stiff broom or the use of an industrial vacuum cleaner is usually required.

If the surface of the concrete is relatively dry, the flame may pass over it without scaling. This problem can generally be overcome by wetting the concrete over-night.

Apart from texturing the concrete, the flame technique can be used to remove paint and similar thin layers.

Concrete can be cut by the use of a thermic lance, special saws, and stitch drilling with diamond drills.

Bibliography

American Concrete Institute. Guide to joint sealants for concrete structures. Title 67–31. *J. ACI*, July 1970, pp. 489–536.

Canadian Portland Cement Association. *Design and Control of Concrete Mixtures.* Portland Cement Association, Skokie, USA, 1978, 131 pp.

Concrete Society. *Symposium on Design for Movement in Buildings, London*, 14 October 1969. Five papers.

Deacon, R. C. *Watertight Concrete Construction.* Code 46.504, Cement and Concrete Association, London, 1978, 30 pp.

Harrison, T. A. *Tables of Minimum Striking Times for Soffit and Vertical Formwork.* CIRIA Report No. 67, October 1977, 24 pp.

Harrison, T. A. *Formwork Striking Times—Methods of Assessment.* CIRIA Report No. 73, October 1977, 38 pp.

Hoff, C. H. and Houston, B. J. Non-metallic water stops. Title 70–2. *J. ACI*, January 1973, pp. 7–13.

Hughes, B. P. *Elimination of Shrinkage and Thermal Cracking in Water Retaining Structures.* Technical Note 36, Construction Industry Research & Information Association, London, December 1971, 22 pp.

Johansson, L. *Flame Cleaning of Concrete.* Report 4:77, Swedish Cement and Concrete Research Institute, Stockholm, 1977, 20 pp.

Johns, H. Joint sealing in reclamation canals. Title 70–3, *J. ACI*, January 1973, pp. 14–18.

Monks, W. L. *The Performance of Water Stops in Movement Joints.* Code 42.475, Cement and Concrete Association, London, October 1972, 8 pp.

Murphy, F. G. *The Effect of Initial Rusting on the Bond Performance of Reinforcement.* CIRIA Report No. 71, November 1977, 28 pp.

Perkins, P. H. *Concrete Structures: Repair, Waterproofing and Protection*. Applied Science Publishers, London, 1977, 302 pp.

Perkins, P. H. *Swimming Pools*, Second Edition. Applied Science Publishers, London, 1978, 398 pp.

Pink, A. *Winter Concreting*. Code 45.007, Cement and Concrete Association, London, 1978, 28 pp.

Reynolds, C. and Steedman, J. C. *Reinforced Concrete Designers Handbook*, Eighth Edition. Ref. 12.074, Cement and Concrete Association, London, 1976, 439 pp.

Shirley, D. E. *Hot Weather Concreting*. Code 45.013, Cement and Concrete Association, London, 1978.

Turton, C. D. To crack or not to crack. *Concrete*, November 1974, 5 pp.

Wright, P. J. F. *The Design of Joint Sealing Grooves in Relation to Joint Spacing and the Problem of Sealing*. Laboratory Note LN/474/PJFW, Road Research Laboratory, January 1964, 15 pp. (not published).

CHAPTER 8

REINFORCED CONCRETE LIQUID RETAINING STRUCTURES CONSTRUCTED ON OR IN THE GROUND

8.1. *Introduction*

It is rarely that designers of liquid retaining structures have complete freedom in the selection of the site for the structure. In the case of retaining structures for water supply, it is usually the availability of the ground and its elevation which control the choice. With sewerage structures it is distance from inhabited areas, location of a suitable outfall and the need to keep pumping costs to a practical minimum which are likely to be the determining factors.

This chapter considers various aspects of construction as they affect work on site.

It is obvious that careful preplanning is needed for a project of any size. Many sites use ready-mixed concrete instead of having a batching plant on site. This can be quite satisfactory provided there is prior discussion between the contractor, the supplier and the resident engineer, to ensure that all the requirements of the specification are met and that the rate of delivery of the concrete satisfies the job requirements.

Good site organization and good relations between the resident engineer and the contractor's agent are essential for smooth running of the contract and completion of a satisfactory job. Recently there has been a move towards the use of plain (unreinforced) concrete for the walls of large liquid retaining structures such as reservoirs. This is discussed in some detail in Chapter 9.

Code of Practice BS 5337, Clause 4.9, sets out 'degrees of exposure' into three classes, namely A, B and C, and Table 8 in the Code lays down minimum cement contents for the concrete used in two of these classes, A and B, to ensure impermeability and durability.

Figure 8.1 shows these requirements for a typical roofed water retaining structure. The authors do not agree that 'moist conditions' can necessarily be equated with 'corrosive conditions'. The implications of the Code are that if the water level is as shown in Fig. 8.1, the walls and columns would

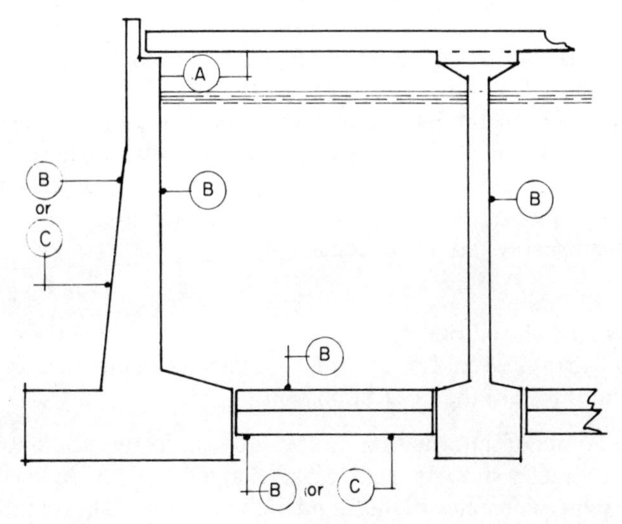

Fig. 8.1. *Diagram showing degrees of exposure—BS 5337, Cl. 4.9.*

have to comply with exposure condition 'A' above normal top water level, and exposure 'B' below this level. As it is not practical to have two different concrete mixes in the same structural unit, this would necessitate concrete for exposure 'A' for both walls and columns. However, if the tank were normally filled to the underside of the roof, then walls and columns could be constructed in concrete to meet exposure condition 'B'.

Because of the authors' reluctance to accept as technically sound this approach to concrete mixes for liquid retaining structures, some of the recommended cement contents in this book do not agree completely with the Code. Readers of the book will of course decide themselves which recommendations they follow.

Detailed comments on the Code are given in Chapter 11. Generally, the recommended cement contents are slightly higher than those in the Code.

8.2. *Sub-soil Surveys and Interpretation of Results*

General Considerations

It is now normal practice in this country to carry out sub-soil investigations (also called sub-soil surveys) in the early part of the design stage of all building and civil engineering projects of any significance.

The work should only be entrusted to experienced firms and they should be asked to give practical interpretation and advice on the results of their survey. The theory and practice of sub-soil investigations is outside the scope of this book and readers should refer to the bibliography at the end of the chapter. The work falls within the province of soil mechanics, particularly the methods of sampling and the laboratory testing of the soil. While each case must obviously be treated on its merits, the authors hope that the notes which follow will form a guide for the practical and realistic solution of some of the more general problems which arise when the results of sub-soil surveys are being considered.

Reasons for Sub-soil Surveys

There are many reasons for carrying out sub-soil investigations of which the following three are the most important:

(a) To obtain information on the sub-soil, its physical and chemical characteristics, to enable the designer to decide on the type, dimensions, etc., of the foundations and other parts of the structure below ground level.

(b) To ascertain whether the sub-soil and/or ground water is likely to be aggressive to the concrete in contact with it.

(c) To obtain sufficient information for the contractor to appreciate the problems involved in carrying out work below ground level and to price his contract accordingly.

It is not intended to give information and advice on methods, etc., for carrying out site investigations as there is adequate published material already available; readers should consult the bibliography at the end of this chapter for some of the more important documents which include CP 2001—Site Investigations, and CP 2004—Foundations. The authors would however like to emphasize the need for a sufficient number of boreholes of adequate depth to give as clear a picture as possible of sub-soil conditions. The cost of a few extra boreholes is negligible compared with cost of changing the design of foundations during the contract, due to unexpected variations in sub-soil conditions.

Interpretation of Results

In almost all sub-soil investigations, with the exception of rock, the pH of the soil and the ground water (if any), should be given; where the pH is below 6·5, a chemical analysis should be carried out to ascertain the type and concentration of the acid compounds present.

The following factors must be kept in mind and taken into account:

(a) There are many practical difficulties in obtaining truly representative samples of soil and ground water and in carrying out the analysis.

(b) All figures should be viewed with caution and considered with all other relevant information.

(c) Due to excavation and sub-soil drainage, including cut-off drains, sub-surface conditions may change appreciably during the construction period and the lifetime of the structure. This can result in significant changes in the chemical composition of the ground water in contact with the structure. For example, aggressive ground water may be practically static, i.e. the velocity of flow may be extremely slow, as in clay soils, or may be appreciable, as in gravels. Where the aggressive chemicals come from a specific deposit (e.g. an industrial tip), their quantity is obviously limited and they may be gradually leached from the soil.

(d) The degree of exposure to attack by aggressive chemicals should be assessed in the site investigation report. The following conditions are listed in increasing severity of attack:

    (i) dry ground conditions;
    (ii) stagnant water (practically no flow);
    (iii) flowing water.

    Seasonal fluctuations of ground water level and its chemical composition, as well as its direction of flow and velocity can have an important bearing on the severity of attack.

(e) Aggressive chemicals in a dry condition are unlikely to attack good quality, well compacted concrete. Therefore if it is practical to lower the water table level permanently and thus provide good sub-soil drainage conditions, the danger of attack on the concrete will be significantly reduced. In fact it is unlikely, except in the case of industrial tips, that dense, impermeable concrete will be attacked in well drained soils above water table level under climatic conditions existing in this country.

    Methods of protecting concrete and other structural materials in an aggressive environment have been dealt in Chapter 6.

8.3. *Foundations*

General Principles

The majority of foundations for liquid retaining structures come within the following categories:

(a) Mass concrete and reinforced concrete strip footings and column bases.
(b) Reinforced concrete rafts uniformly supported on the ground.
(c) Reinforced concrete rafts supported on piles. There are various types of piles, to which specialist piling firms give their own trade names. Essentially, these can be considered to fall into three groups:
   (i) precast concrete piles;
   (ii) bored *in situ* concrete piles;
   (iii) Driven or bored shell piles.

   All of the above can be either end-bearing piles, friction piles, or a combination of end-bearing and friction.

   The decision as to which pile to use requires detailed consideration in the light of sub-soil investigations, site conditions including access, and proximity of other existing structures, and the type of structure to be supported.

All the above types of foundations, including the piles, are dealt with in detail in a number of text books to which the reader is referred. However, the authors' experience suggests that bored *in situ* concrete piles present a number of special problems from the site construction aspect and therefore the following notes are given.

Bored piles constructed of *in situ* concrete can vary in dimensions from about 500-mm diameter and 6·00 m long to 2·00-m diameter and 35·00 m long. They have one feature in common, namely that the concrete must be self compacting. This means a high slump, at least 125 mm and it may well be 150 mm. If the concrete has to be placed with a tremie pipe under water, then it is usual to specify a 'collapse' slump of 150–200 mm. For durability, the concrete should have a high cement content and as low a water/cement ratio as possible. This will generally mean that a plasticizer has to be used. If the pile is reinforced, then a chloride-free admixture may be needed.

In some cases the ground and/or ground water may be aggressive to Portland cement concrete and it is then particularly important that the

Fig. 8.2. *Large service reservoir: view of diaphragm wall prior to casting facing concrete. (Consulting Engineers: John Taylor & Sons; Contractors: Costain Civil Engineering. Courtesy: Colne Valley Water Co., Rickmondsworth.)*

concrete should possess maximum durability characteristics. The authors consider that every effort should be made to keep the water/cement ratio to 0·50.

This type of pile is usually cased and the casing is normally withdrawn as the concrete is placed which means that the outside of the pile is in contact with the sub-soil and ground water while it is still plastic and is therefore specially vulnerable to chemical attack and dilution of the cement paste by ground water. It is obvious that from the point of view of durability, the various proprietary types of pile in which a permanent shell or casing is used, have a definite advantage. In some projects diaphragm walls are cast first and then the structural (final) wall is cast against this after the excavation is complete. Figure 8.2 shows such a wall.

8.4. *Underdrainage of Site*

General Considerations

Underdrainage of sites is carried out for two main reasons:

(a) To prevent the build-up of ground water pressure which can result in uplift (flotation) and cause serious damage to the structure.
(b) To monitor any leakage which may develop in the floor.

It is generally correct to say that reason (b) is given a much higher priority with water engineers than with designers of sewage tanks.

Materials and Layout

The pipes used for sub-soil drainage are

(a) porous concrete to BS 1194;
(b) perforated dense concrete, for which there is as yet no British Standard, but there is the US Standard C 444–68—*ASTM Standard Specification for Perforated Concrete Pipe*;
(c) plain dense concrete pipes to BS 4101 or BS 556, laid with open joints;
(d) unglazed clayware pipes to BS 1196, with plain butt joints;
(e) clayware pipes to BS 65, laid with open joints;
(f) slotted plastic pipes, for which at present there is no British Standard;
(g) asbestos cement pipes to BS 3656, either laid with open joints or with the barrel specially perforated; the perforations are not covered by the Standard.

Of the above seven types of pipes, the only one in which the rate of infiltration is laid down in a national standard is porous concrete (BS 1194). This Standard covers two classes of pipes, the classification being related to the external load carrying capacity. Infiltration rates, which vary directly with the diameter, are also included in the Standard.

If the sub-soil or ground water is aggressive, it is recommended that porous concrete pipes should not be used as the porosity makes them particularly vulnerable to attack.

To facilitate the admission of ground water to the drainage system and to help prevent silting up of the points of entry, it is recommended that a properly graded filter be placed around the pipes. The filter should be clean granular material, graded from 20 mm down to 5 mm.

The layout of the underdrainage system and the diameter of the pipes will depend on site conditions. On some sites the provision of a simple perimeter drain 100- or 150-mm diameter would be adequate to keep the site reasonably dry. On other sites, particularly if leakage is to be checked during the life of the structure, a complete system of under drains would be required. For the purpose of checking leakage it is important that the whole of the area below the floor should be divided into a number of separate drainage areas, each of which is connected to the main collector drains at manholes. Two suggested layouts are shown in Figs. 8.3 and 8.4.

Flotation

The problem of flotation seldom arises with water reservoirs principally because there is no hydraulic reason for sinking them in the ground, and they are usually constructed on high-level sites rather than low-level sites.

With sewage tanks, however, the reverse is the case; if the sumps of pumping stations are considered as liquid retaining structures, then these may be at an appreciable depth below water table level. In such cases, the problem of uplift and flotation can assume serious proportions. The simplest way of overcoming the tendency of the tank to 'float' as the result of the upward pressure of ground water, is to increase the dead weight of the structure. Concrete is a comparatively heavy material, the specific gravity being about 2·4. The upward pressure of the ground water on the floor slab is resisted by the dead weight of the tank, the friction between the tank walls and the ground and, when the tank is full, by the weight of its contents. The trouble arises when tanks are under construction at the stage when the floor has been cast and the walls partly cast. The ground water in such cases is controlled by various ways, the most common being well points, sumps, wells and boreholes. If at a critical point during construction the pumping plant breaks down, the whole structure (as far as it has been built) may be tilted and/or lifted and break its back. When such site conditions exist, continuous supervision and stand-by pumping plant are essential.

A useful safety device is pressure-relief valves; these are fixed in the floor and operate at a predetermined pressure from the ground water. They can be sealed-in when the structure is complete. In some cases they are left in as a permanent safety feature; and example is gunite swimming pools in which they are almost invariably installed.

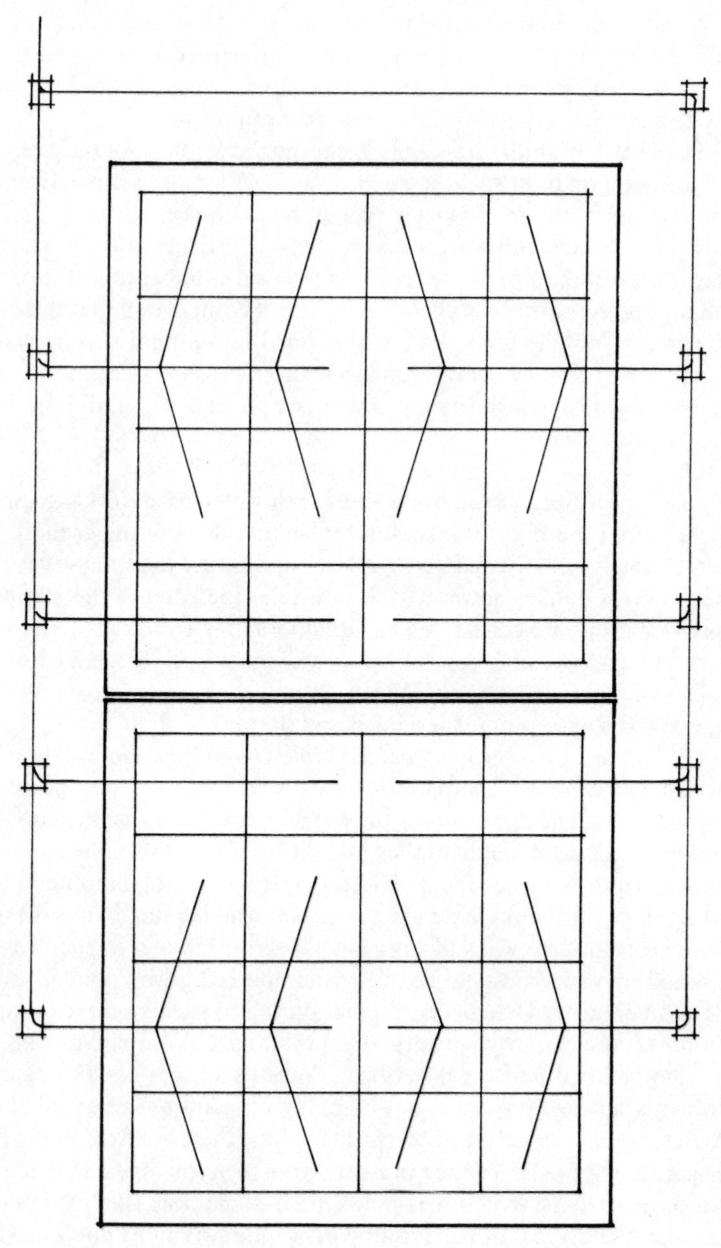

Fig. 8.3. *Underdrainage of reservoir, alternative layout.*

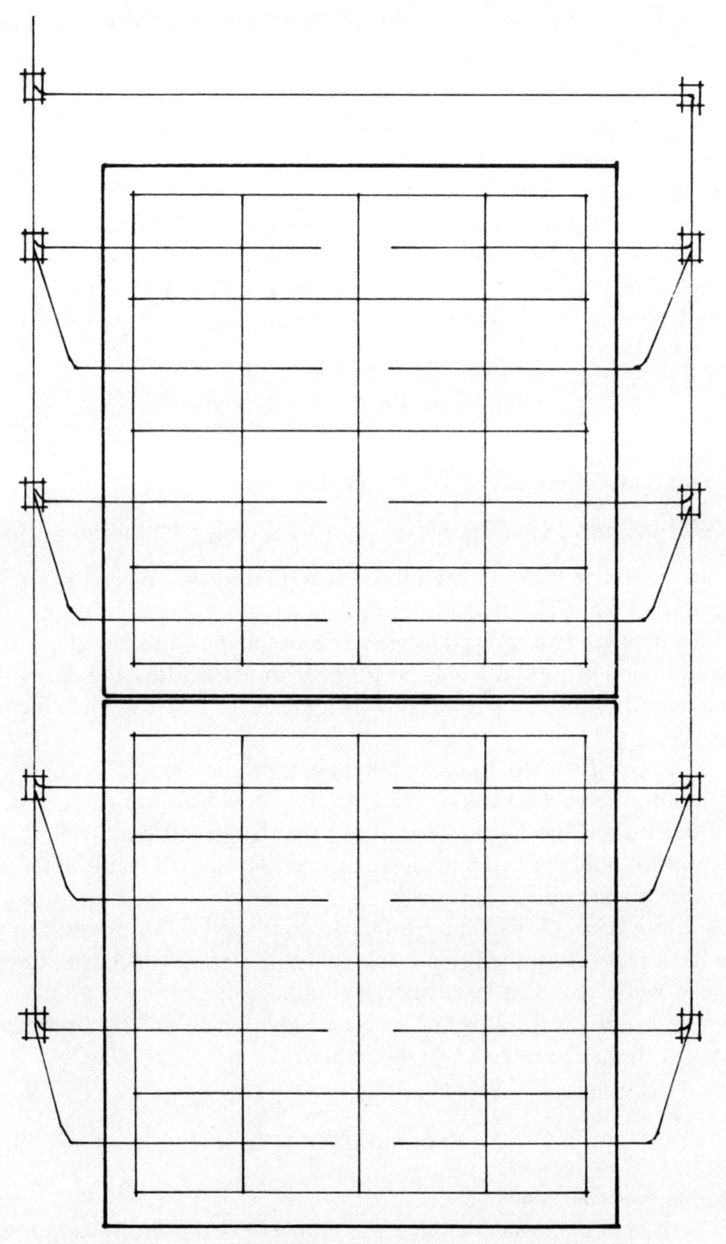

Fig. 8.4. *Underdrainage of reservoir, alternative layout.*

The following simple calculation shows how vulnerable a deep tank can be to flotation.

Tank: 10·00 m × 10·00 m on plan, 6·00 m deep.
Ground water level: 5·00 m above underside of floor of tank.
Walls: 0·50 m thick.
Floor: 0·30 m thick.
Area of underside of floor: 11·0 m × 11·0 m = 121 m^2.
Unit water pressure on underside of floor: 5000 kg/m^2.
Total water pressure: $(5 \times 10^3) \times (1{\cdot}21 \times 10^2) = 6{\cdot}05 \times 10^5$ kg.
Dead weight of empty structure with walls and floor completed:

$$\begin{aligned} \text{Floor: } 121 \times 0{\cdot}30 &= 36{\cdot}30\,\text{m}^3 \\ \text{Walls: } 40 \times 6 \times 0{\cdot}5 &= \underline{120{\cdot}00\,\text{m}^3} \\ & 156{\cdot}30\,\text{m}^3 \end{aligned}$$

Assuming concrete weighs 2400 kg/m^3

$$\text{Total weight of concrete} = (1{\cdot}56 \times 10^2) \times (2{\cdot}4 \times 10^3) = 3{\cdot}8 \times 10^5\,\text{kg}$$

From the above it can be seen that the weight of the completed structure (empty) is only 60 % of the total upthrust from the ground water. While there is likely to be appreciable friction between the outside of the tank and the adjoining ground it would be prudent to neglect this and design for a total nett uplift of $2{\cdot}2 \times 10^5$ kg to which should be added a safety factor of, say, 25 %.

As previously stated, the dead weight of the structure can be increased by increasing the wall and floor thickness and this will give some saving in reinforcement. Another method is to provide tension piles which will be designed to hold the structure down against the upward thrust of the water pressure on the floor.

Another method, which is in fact seldom used, is to provide sub-soil drainage pipes connected to a small pumping station which will lower the ground water to a safe level when it is required to empty the tank. Some form of automatic warning system is desirable which will operate when the water in the tank reaches a predetermined level.

8.5. *Floors*

General Considerations

The type of floor considered here is assumed to be uniformly supported on the ground and is not a suspended slab.

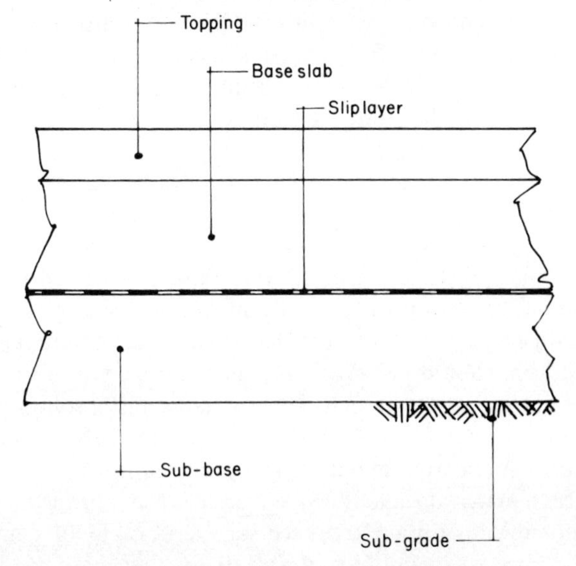

Fig. 8.5. *Illustration of definitions.*

There are no generally accepted principles of structural design for reinforced concrete slabs uniformly supported on the ground in the same way as there are for suspended slabs.

Ground floor slabs of industrial buildings can be satisfactorily designed on the basis of road slabs, and for such slabs the design criteria laid down in Road Note 29 and the SI version of axle loads in a Technical Note of 1978, are often used.

For liquid retaining structures a different design approach is used. An example of the design of structural floor slabs is given in Part 1 of this book, from which it will be seen that the reinforcement is provided mainly to control thermal contraction cracking. While some of the expressions used in this section are quite familiar to road engineers they may not be in common use with other engineers, and therefore the following definitions are given:

Sub-grade: The material immediately below the sub-base, consisting of the natural sub-soil or imported selected fill.

Sub-base: A layer of material between the sub-grade and the base slab, such as site concrete.

Base slab: This is also referred to as the structural floor slab.

These definitions are illustrated in Fig. 8.5.

This section is intended to deal with the floor slab proper, as in many liquid retaining structures the perimeter bays of the floor form the foundation for the wall panels, and consequently are appreciably thicker than the remainder of the floor. In addition, the wall reinforcement is anchored into these perimeter slabs.

Sub-grade

It is unusual for the floors of liquid retaining structures to be founded on other than the natural (virgin) ground. If imported fill has to be used, as may be the case where the upper layers of soil are aggressive to concrete, for example an industrial tip, and therefore have to be removed and replaced by inert material, care should be taken to ensure compaction of the new fill. However, the authors consider that the most satisfactory solution in such cases is to ignore the imported fill and design the floor slab as a suspended slab, the vertical supporting members being carried down to the natural ground. Unless a material such as lean concrete is used for the new fill it is virtually impossible to obtain the necessary uniformity of compaction which will ensure that no subsequent differential settlement of the floor slab will occur.

If hardcore is used to make up the levels, then this should be broken up to a maximum size of 50 mm. While many types of hardcore are suitable, material containing sulphur compounds should not be used as these compounds may be present as sulphates or acids, which under damp conditions may go into solution and attack the concrete. If there is any doubt about the chemical composition of the hardcore, a chemical analysis should be carried out. The use of sulphate resisting Portland cement in the sub-base concrete will be adequate safeguard against sulphate attack under most conditions but it will give little or no advantage if acids are present.

Alternatives to hardcore are

(a) cement stabilized granular material; under favourable circumstances (such as where a suitable free-dug material is close by), this may be technically and economically more satisfactory than hardcore;
(b) lean concrete ('wet' lean concrete or 'dry' lean concrete).

For detailed information on the material (a), reference should be made to the bibliography at the end of this chapter.

Lean concrete, which can be obtained from most ready-mixed concrete firms, is more likely to be used than cement stabilized material and therefore brief recommendations on this material are given below.

Lean concrete is intended to have a low compressive strength, 10–14 N/mm^2, and to be slightly flexible compared with structural (paving) quality concrete. It is used in large quantities as a sub-base for roads. The mix proportions are in the range of 16:1 to 24:1 (aggregate to cement). Lean concrete is batched and mixed in the same way as ordinary concrete and the aggregates should comply with BS 882—*Aggregates from Natural Sources for Concrete*, although air-cooled blastfurnace slag to BS 1047 can be used as the coarse aggregate.

If the maximum size of coarse aggregate is 20 mm, the sand content should be 40% of the total weight of aggregate.

With an aggregate/cement ratio of 20, which gives a cement content of about 120 kg/m^3, the dry density would be about 220 kg/m^2 and the average compressive strength at 28 days would be about 14 N/mm^2.

Proper compaction is essential, and with 'dry' lean concrete this is best achieved by rolling, using either an ordinary road roller or a vibrating roller. For comparatively small areas a vibrating plate can be used. The concrete is spread in layers within the range of 100–250-mm thickness and each layer is compacted.

Wet lean concrete is quite workable, and would have mix proportions of about 1:16 (cement to total aggregate) and a water/cement ratio of about 0·90. It is likely to be reasonably self-compacting and once it has hardened no settlement within the concrete will occur.

Sub-base

This normally consists of 'site' concrete, varying in thickness from 75 mm to 150 mm, depending on the quality of the sub-grade, and the weight of contractor's plant which will use the slab.

The purpose of site concrete is to provide a clean, level and firm base on which to lay the structural slab. Some experienced engineers use the same mix proportions for the site concrete as for the structural slab, while others use a leaner mix. It is easier to obtain a good smooth finish to the concrete when the higher quality mix is used. For the leaner mix the authors consider that mix proportions of 1 part ordinary Portland cement to 8 parts of fine and coarse aggregate by weight can be used.

The concrete should be laid in bays and compacted in the usual way, but special attention should be paid to the surface finish so that this is as smooth as practical; the finish obtained with a wood float is satisfactory.

The concrete sub-base is usually laid without reinforcement and the bays are arranged so that they do not coincide with the bays in the structural slab which will be laid on top. Unless a full movement joint is provided right across the structure, all the joints in the sub-base slab are plain butt joints, without water bars and without sealing grooves.

After compaction and finishing, the concrete should be cured by covering with 500-gauge polyethylene sheets, well lapped and held down around the edges with boards and concrete blocks. A curing period of 4 days is adequate except in cold weather when it may have to be extended to at least 7 days.

Recommendations have been given in Section 8.4 on underdrainage of sites, but the majority of structures are constructed 'in the dry' even though the site may have a high water table. This is achieved by the use of well point de-watering or sumps and pumps. The need to keep the ground water level below foundation and floor level during early stages of construction in order to avoid uplift has already been dealt with. However, even if there is no danger of uplift, it is important that the water should not rise so as to damage the concrete while it is maturing.

In Chapter 6, information was given on factors affecting permeability and it was stated that good quality concrete can be assumed to become impermeable after about 15 days from date of casting. If there is water pressure on the base during the early life of the concrete, say within the first 7 days, then permanent damage can be caused.

Sliding layer

A sliding or slip layer should be provided between the sub-base and the structural slab. This is to reduce friction between the sub-base and the underside of the base slab which is important during the early life of the concrete. The materials used for this slip layer include the following:

(a) Polythene sheets, 1000 gauge, in single or double layer.
(b) Reinforced building paper.
(c) Roofing felt in single or double layer.
(d) Two or more coats of bituminous emulsion, with or without hessian or fibreglass mesh embedded between the coats.

It should be emphasized that this layer is not a waterproof membrane although the materials used are in fact the same or similar to those used for vapourproof membranes in concrete ground floor construction for houses.

A good job can be obtained by using a proprietary polythene sheet (such as 'Bitu-thene), which has a bitumen adhesive on the underside. The advantage of this material is that it bonds to the base concrete and can therefore be laid without 'rucking'.

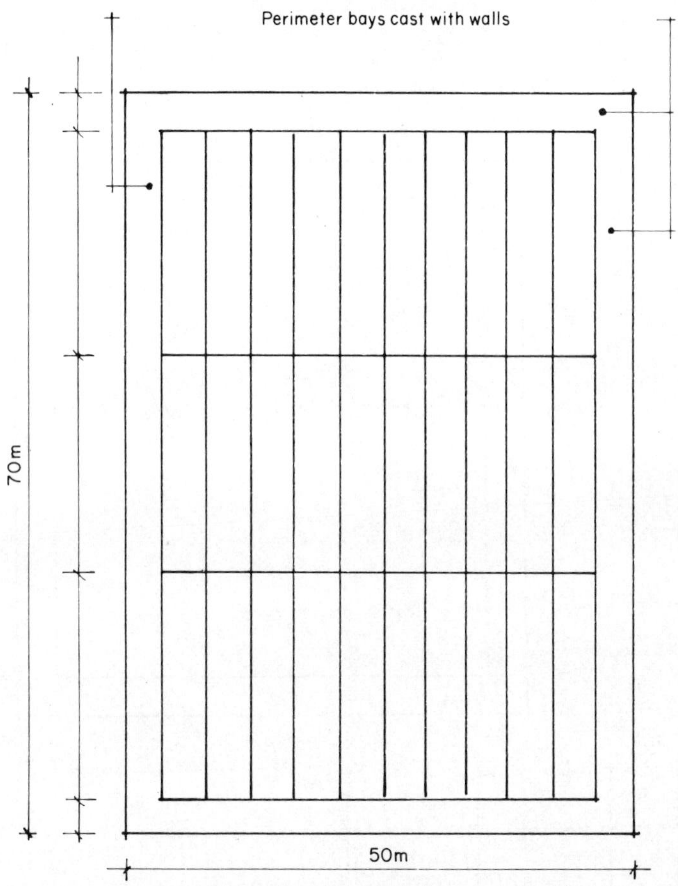

Fig. 8.6. *Bay layout for open reservoir floor.*

The Structural Slab

General Considerations

As a general rule, leakage through floor slabs is more difficult to locate and to remedy than leakage through the walls. Therefore it is worth while to take special care over all details of construction, particularly the joints. When leakage does occur it is usually through the joints and therefore the shorter the total length of joints in the floor, the smaller the chance of leakage. The bay layout will depend on the shape of the structure, (square, rectangular or circular) and whether there is a roof supported on columns. Figures 8.6 and 8.7 show typical bay layouts for the floors and slabs of rectangular, open and roofed reservoirs.

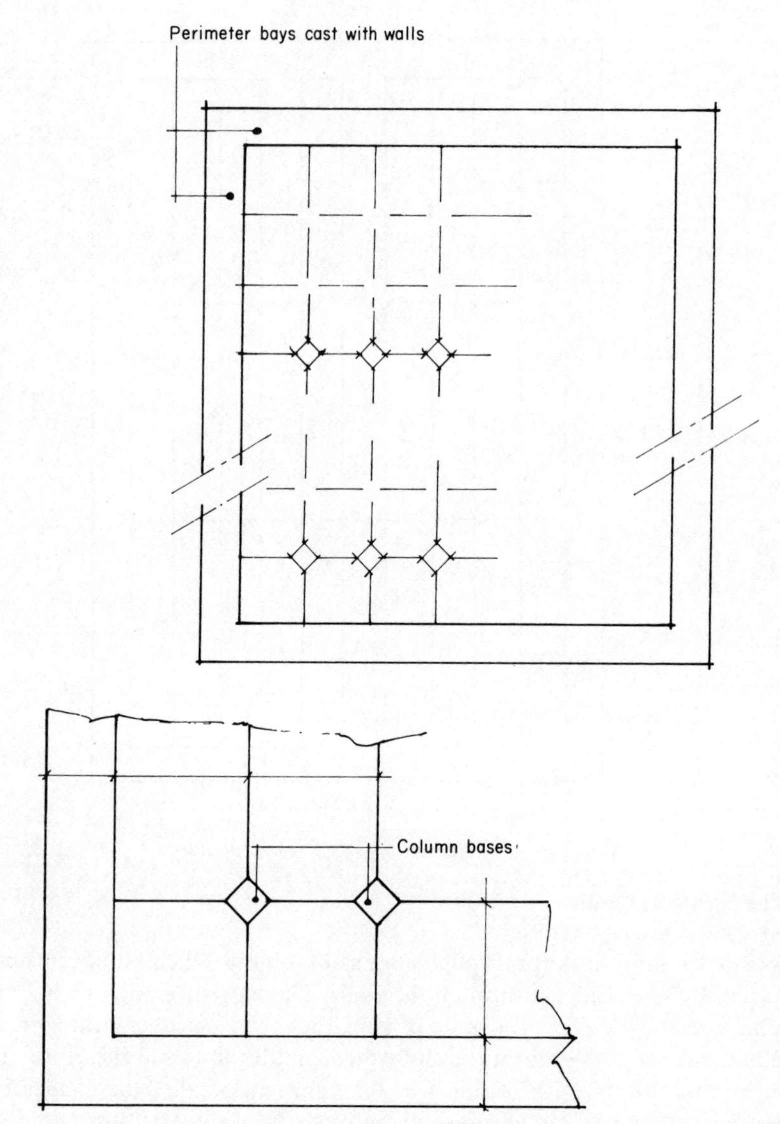

Fig. 8.7. *Bay layout for closed reservoir floor.*

Mix Proportions

The remarks in the Introduction at the beginning of this chapter on Code requirements for mix proportions and the authors' reasons for disagreeing with them should be referred to. In the majority of structures where the floor slab is uniformly supported on the ground, the thickness of the base slab is likely to be in the range 200–500 mm. Therefore the question of special mix design to reduce heat of hydration and thus thermal contraction stresses will not arise.

Cement: Ordinary Portland cement

Minimum cement content: For slabs up to 500 mm thick, 360 kg/m^3 of compacted concrete. For slabs thicker than 300 mm the cement content can be reduced to 330 kg/m^3.

Aggregates: To comply with BS 882, and 20-mm maximum size. There is no technical objection to 40-mm size provided there is adequate distance between the reinforcing bars. With the larger aggregate size there should be increased workability with the same water/cement ratio.

Slump: 50 mm $\pm$ 25 mm

Water/cement ratio: Maximum 'free' water/cement ratio should be 0·50.

Admixtures: These should only be permitted with the written approval of the resident engineer. Admixtures containing calcium chloride should not be used. Some authorities claim that air entrained concrete is more impermeable than ordinary concrete. The use of an air entraining admixture so as to give an air content of $4\frac{1}{2}\% \pm 1\frac{1}{2}\%$ will improve both workability and cohesiveness and greatly increase frost resistance in exposed positions.

It should be noted that there is nothing absolute about the cement contents recommended. If they are included in the specification, and site conditions such as high summer ambient temperatures, or use of timber formwork, etc., result in undesirably high temperature in the maturing concrete, then other factors being satisfactory, the cement content may be reduced.

Strength

A requirement for 'characteristic' strength can be included as this is relatively easy to measure. However, as mentioned in the section on mix design in Chapter 6 (p. 157) this specified characteristic strength should be compatible with the type of cement and aggregates used, and with the specified cement content and water/cement ratio and/or slump. For the quality of concrete recommended in this book, the characteristic strength is likely to be significantly higher than that required by the design criteria. Trial mixes are strongly recommended as these will help to ensure that the specification is understood by all concerned and that the required quality of the concrete is achieved, and will help prevent disputes on site.

Reinforcement

The detailing of reinforcement is considered in Part 1; for the purpose of this section it will be assumed that one layer of reinforcement is used and that this is placed in the top of the slab with a minimum cover of 40 mm, although many engineers provide 50-mm cover in floor slabs. The reinforcement can be placed in position and supported on stools before any concrete is cast (this is the procedure when a double layer is used). An alternative, which in the opinion of the authors is quite acceptable, is to place and compact the concrete up to the required level, so that the reinforcement will have the necessary cover; then immediately place the reinforcement which has been previously fabricated into a mat or mats of the correct bay dimensions. The concrete topping is then placed, compacted and finished. It is essential that the work in each bay should be carried out as one continuous operation. This requires good site organization.

For the reasons given in Chapter 7, the authors do not recommend that the reinforcement is carried across the bay joints unless there is a sound structural reason to do so.

Bay Layout and Casting Sequence

The bay layout will depend on the shape of the structure, the presence or otherwise of columns and the weight of reinforcement in the floor slab.

Figure 8.7 shows a bay layout in a rectangular structure which contains columns. If this is compared with that shown in Fig. 8.6 it will be seen that the structure without the columns has a much simpler bay layout.

In each case the perimeter bays form the foundations for the walls and therefore these are usually cast before the main floor slab. However, there are many advantages if the oversite concrete is cast first provided this is of adequate strength to carry construction traffic and the stacking of materials. Using the layout shown in Fig. 8.6, the floor can be constructed in strips 4·3 m wide with stress-relief joints at 21-m centres, which would divide the 63-m length into three bays. The 63 m can, if desired, be cast in one continuous operation, provided the slab thickness, and weight of reinforcement to control cracking, are adequate. A slab thickness of 250 mm would be suitable. A suggested casting sequence for these bays is shown in Fig. 8.8. This is based on the assumption that one 63-m-long strip

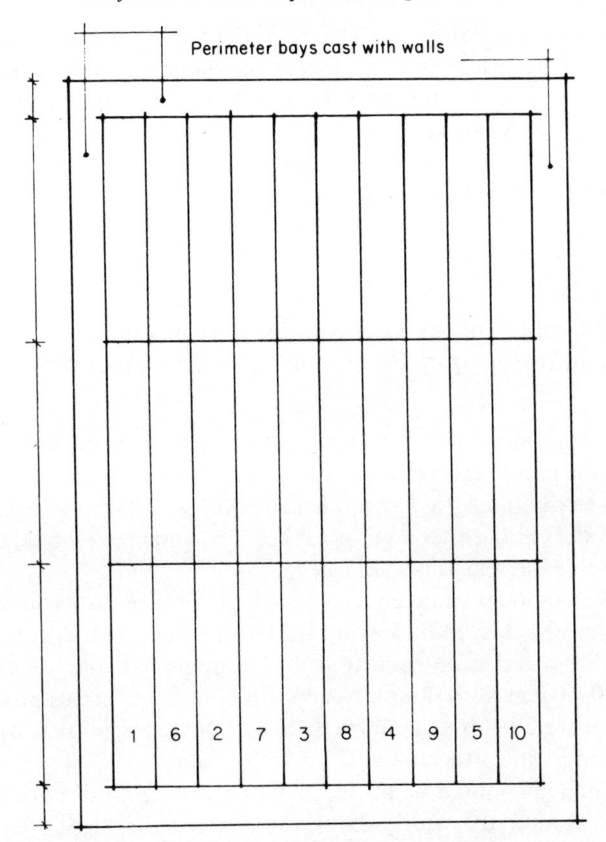

Fig. 8.8. *Casting sequence for floor bays—open reservoir.*

would be cast each day. When infill bays are cast, the vibrating tamper works off the two bays on each side, and the concrete in these bays must be mature enough not to be damaged by the tamper.

In a structure with a roof supported on columns, the spacing of the columns generally determines the bay layout, as shown in Fig. 8.7. Some open structures are of considerable size and in such cases it is worth while to give special consideration to the bay dimensions and layout with a view to reducing as far as possible the length of the joints. Joints are expensive and are the most vulnerable part of the floor from the point of view of leakage. Leakage through a floor is the contractor's nightmare, as it is very difficult to locate and equally difficult to remedy. As an example, take an open structure in which the main floor, excluding the perimeter bays which carry the walls, measures 90·0 m × 180·0 m. If the bays were laid out on a 6·0 m × 6·0 m square grid, the total length of joints would be about 5400 m (5·4 km). With rectangular bays 6·0 m × 20·0 m, the total length of joints would be about 3500 m (3·5 km) a saving of some 1900 m (1·9 km). If the bays were 10 m × 10 m the total length would be about 3·2 km, a saving of 2·2 km. However, with the longer bays, a heavier weight of reinforcement would be required to control cracking.

Joints

The whole problem of joints has been discussed in some detail in Chapter 7; based on the reasoning in that chapter the following recommendations are made:

(a) Longitudinal and transverse joints between bays should be plain butt joints, except where they are formed with crack inducers.

(b) The whole length of the joint should be sealed with an external type water bar of either PVC or rubber. The water bars should be solvent welded at the joints (Fig. 8.9).

(c) Steps should be taken to reduce bond between adjacent bays; this can be achieved by leaving the concrete as it comes from the side forms and stop-ends and by stopping off the reinforcement 40–50 mm from the joint. A coat of bituminous emulsion on the old concrete before the new concrete is placed against it will reduce bond still further.

(d) The joint should be provided with a sealing groove about 20 mm × 20 mm; the width should not exceed the depth. The groove should be temporarily filled with an inert material which will not adhere to the concrete. As late in the construction process as possible, the groove should be cleaned out and sealed with a high-grade sealant. For information on joint sealants the reader is referred to Chapter 5.

(e) The design of the joint should be as simple as possible; in the vast majority of cases there is no need for joggles or shear keys as these serve no useful purpose and merely interfere with the compaction of the concrete and increase cost.

Compaction and Finishing

No matter how high the quality of the concrete is when it leaves the mixer it must be transported and placed without segregation occurring; it must then be compacted and cured to ensure impermeability and strength.

Fig. 8.9. *Floor of sewage tank showing reinforcement and external type PVC water bars laid out on site concrete. (Courtesy: Servicised Ltd, London.)*

Fig. 8.10. *Heavy reinforcement in floor of sewage settling tank.*

Slabs up to about 200 mm in thickness, can be satisfactorily compacted by vibrating beam tampers, provided the reinforcement is not congested. For slabs up to 200 mm thick, containing a single layer of reinforcement, hand operated vibrating tampers working off the side forms will provide adequate compaction, except around the edges of the bay adjacent to the side forms where poker vibrators should be used to supplement the tamper.

If a double layer of reinforcement is used, then the authors recommend poker vibrators followed by a vibrating beam. When the main compactive effort is obtained from tampers, it is necessary to provide a surcharge on the concrete and this is usually 1/5 of the thickness of the slab. Figure 8.10 shows a heavily reinforced floor of a sewage tank.

To obtain a true, smooth surface to the compacted concrete, the finishing techniques used for industrial floor slabs can be adopted. The actual method used will depend on site conditions and the standard of finish required. For large, level floors (as in water reservoirs), power floats, if properly operated, will give a very good finish. It is generally accepted that the minimum area on which it is economic to use such power tools is about $100\,m^2$.

The terms power trowel and power float are interchangeable and there are two basic types, one incorporating a revolving steel disc and the other revolving blades. Some experienced floor layers consider that the best results are obtained by using the disc trowel first followed by the blade trowel. The most important factor is the experience and skill of the operator. The condition of the concrete, i.e. the finish given by the previous process (beam tamper), and stiffness, are critical for the success of the trowelling operation. If the trowel is used when the concrete is too plastic it will spoil the carefully finished surface of the slab, resulting in waves. On the other hand, if the concrete has been allowed to stiffen too much, the trowelling will be ineffective. The period which should elapse between the completion of the compaction and the start of the power trowelling depends on many factors and must be determined by the operator himself. A simple but useful test is for the operator to walk on the surface and when his boots leave indentations 3–4 mm deep the concrete is likely to be ready for power trowelling.

Figure 8.11 shows a power trowel in operation.

Other methods of finishing the slab include hand trowelling with wood or steel floats, and the use of a bull float or skip float. The two latter tools have long handles and therefore cannot be used in a confined space.

If a particularly smooth, dense surface is required, as is specified in certain sewage treatment tanks, hand trowelling with steel trowels is the best answer.

Curing

Concrete floor slabs must be properly cured if they are to be crack-free, durable and impermeable to water. Curing consists of substantially reducing the rate of evaporation of moisture from the surface of the slab for the first 72–96 h after completion of compaction and finishing. This assists in the hydration of the cement and the blocking of the capillary pores in the cement paste, to which more detailed reference has been made in Chapters 6 and 7. Unless this is done the surface of the concrete is liable to develop cracks and be weak and porous as well. The depth and width of

Fig. 8.11. *Power trowel in operation.*

cracks caused by lack of, or inadequate, curing will depend largely on how much and how quickly the water in the concrete was evaporated from the surface. Generally in this country, the cracks are fine and penetrate about 25–50 mm, but in extreme cases this form of cracking can be so severe as to necessitate the removal and relaying of the slab.

One form of this cracking is known as 'plastic' cracking and is caused by the rapid evaporation of water which 'bleeds' to the surface of the slab while the concrete is still plastic. Plastic cracking has been discussed in detail in Chapter 7.

There are two basic methods of curing the floor slabs of liquid retaining structures: by the use of a curing membrane, and polyethylene sheets.

Curing membranes are produced by numerous firms, each claiming special qualities for their own particular product. The type selected should be resin-based and non-toxic. These compounds can be applied by brush or by spray. They wear off in the course of time, but some residue is likely to remain unless special means are adopted to remove it by high pressure jets, wire brushes, grit blasting, etc. The authors consider that the use of high pressure water jets is most effective as all traces of the membrane are completely washed away. This complete removal is desirable when it is required to bond another layer to the surface of the floor slab. There is no British Standard for curing membranes, but reference can be made to DpT Specification for Road and Bridge Works, 1976, in the section on Curing Membranes for Concrete: Water Retention Efficiency Test. There are also two American Specifications for curing membranes, namely ASTM C 309–73 and C 156.

The use of 500-gauge polyethylene sheets, well lapped and held down around the edges by scaffold boards so that wind cannot blow underneath them, is very effective. Proper curing by one of the methods just described is essential even though the floor may be inside an existing structure, such as reslabbing an existing tank.

Reference should also be made to the sections in Chapter 7 on hot weather and cold weather concreting.

The authors have often encountered the use of 'wet' sand and 'wet' hessian for curing slabs. The danger here is a practical one, the difficulty of ensuring that the sand or hessian is in fact kept permanently wet and is not allowed to dry out during the specified curing period.

It is recommended that in addition to the 'normal' or 'standard' curing methods outlined above, the specification should include a clause requiring the contractor to protect the concrete against solar heat gain during casting. A provisional sum can be allowed for this in the Bill which will be used at the discretion of the resident engineer. The same principle can be adopted for winter concreting techniques. At the time the contract documents are prepared it is impossible to predict the type of weather likely to be experienced during the period of the contract.

Popouts

It sometimes happens, fortunately very rarely, that popouts occur in the exposed surface of concrete slabs. The real cause of these popouts is often difficult to ascertain and sometimes a satisfactory explanation cannot be found. In the UK popouts are attributed to frost action on patina coated gravel particles. The coating is usually very thin, but is highly absorbent. Water slowly penetrates into the surface layer of the concrete and is absorbed by the patina coating; then when freezing conditions occur, the water contained in the patina coating freezes, expands and forces out the piece of aggregate. Figure 8.12 shows such a popout. Another cause is the presence of lignite or coal particles in the coarse or fine aggregate. In many

Fig. 8.12. *View of popout on floor of reservoir.*

such cases the popouts occur at early age, sometimes within 24 h of placing the concrete.

In parts of the world where alkali–aggregate reaction occurs (see Chapter 5), popouts have been found to be due at least in part to this reaction. An investigation in the US was reported in the *J. ACI* of June 1974 (see bibliography at the end of this chapter).

8.6. *Pipework Passing through Floor and Walls*

There are seldom problems when the pipes pass through walls and roof above top water level, but in many cases pipes have to be located below water level and then special care is needed to ensure watertightness. There are two ways of carrying out this work, namely boxing-out and building-(casting-)in as the work proceeds.

Boxing-out

There are differences of opinion as to whether boxing-out or casting-in as the work proceeds is likely to give the better result from a watertightness point of view. The authors favour casting-in whenever this is practicable. There are cases when a hole must be left and the pipe inserted and concreted in later. The two cases most frequently met are when the pipe is not available when the floor/wall is cast, and when the exact position for the pipe has not been finally determined.

It is important that the boxed-out hole should be of adequate size to accommodate the pipe and for the placing and compaction of the mortar or concrete once the pipe is fixed in position. Mortar should only be used around small pipes, up to about 75 mm, but there is no hard and fast rule for this. When mortar is used, the authors recommend that it be gauged with a styrene butadiene latex emulsion as this will help to reduce shrinkage and permeability. The mortar should be as stiff as practical and well compacted.

The outside of the pipe should be clean and free from bitumen, grease, etc. Plastic pipes should have the surface slightly roughened and then immediately before the mortar or concrete is placed, a specially formulated resin bonding agent should be applied to the whole of the outside of the pipe.

The use of a puddle flange located in the centre of the wall or floor slab is not recommended for boxed-out pipes, as it will prove almost impossible to compact the concrete around the flange. However, an external flange can be used as shown in Fig. 8.13. This flange can be fixed after the pipe has been concreted in and can be bedded on a layer of epoxide resin. The detail in Fig. 8.13 can be used for both wall (horizontal pipes) and floor/roof (vertical pipes).

If an external flange is not used, then about 1 month (or longer if possible) after the filling of the annular space around the pipe, a thick coat of styrene butadiene latex grout should be applied over the whole area of the mortar/concrete and extended for at least 100 mm on to the adjacent concrete. This should effectively seal the new mortar/concrete and the perimeter joint with the old concrete. The surface of the new and the old concrete should be wire brushed to remove weak laitence, etc., before the grout is applied. When large diameter pipes have to pass through a floor, roof or wall, it may be necessary to carry the reinforcement across the opening which has been left and this complicates the matter and makes it more difficult to compact the concrete around the pipe.

It is essential that the pipe be securely fixed so that it cannot move when the concrete is placed and compacted. The surface of the hardened concrete

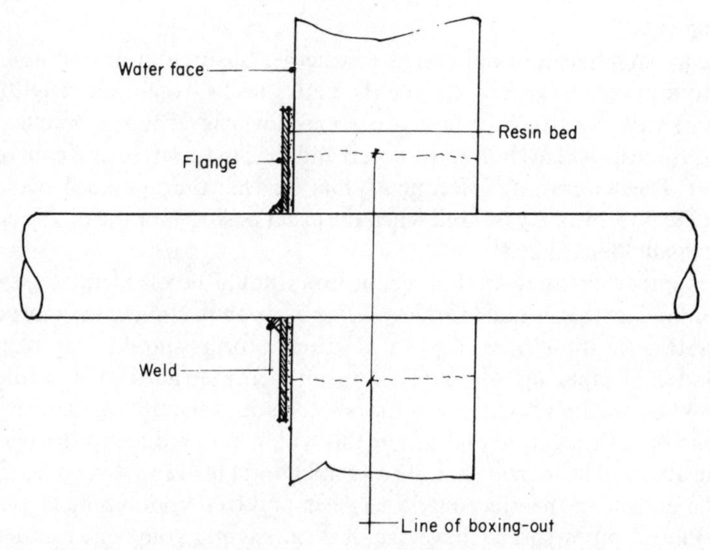

Fig. 8.13. *Boxing-out for pipe through wall or floor.*

within the thickness of the floor, roof or wall must be clean and rough so that the new concrete will bond to it.

Building-in

Unless there are good reasons to the contrary, as mentioned earlier in this section, pipes should be cast-in as the concreting proceeds. In these cases, the provision of a puddle flange is advantageous in increasing watertightness.

An exception to this is when the section (floor, wall or roof) is constructed in gunite or sprayed concrete. In this case, boxing-out should be adopted due to problems with trapped rebound material and difficulty in proper placing behind the flange. There are no rules for the flange diameter, but it is recommended that the flange should be not less than 300 mm larger than the outside diameter of the pipe and for pipes above 900-mm diameter the flange diameter can be increased so that it is 600 mm larger than that of the pipe.

All bituminous and similar coatings should be removed from steel and cast iron pipes and the flanges, as these coatings will interfere with the bond with the concrete. With plastic pipes, a good bond cannot be obtained with the concrete without the use of a specially formulated resin bonding agent which will bond to both the pipe and the concrete. The concrete must be placed while the tack coat of resin is still 'tacky'. An alternative but less effective method is to roughen the outside surface of the pipe.

Because of this problem of bond it is better to insert a short length of ferrous pipe where it passes through the concrete, when this is possible.

When a pipe passes through a floor slab, the provision of a puddle flange at the centre depth of the slab will to some extent interfere with the compaction of the concrete around the pipe. Therefore the authors favour the provision of a flange as shown in Fig. 8.13. If the bedding of the flange is carried out carefully, there should be no leakage around the pipe.

8.7. *Special Problems of Sloping Floors*

For the purpose of this section, a sloping floor is one where the fall longitudinally or transversely exceeds about 1 in 10. Although the heading of this section refers to 'problems', these are more imaginary than real unless the fall exceeds about 35°, i.e. a gradient of 1 in 1·4. For steeper gradients it will be necessary to use a top shutter.

In the case of the intermediate gradients, i.e. between 1 in 10 and 1 in 1·4, the concreting should start from the bottom of the slope and a fairly low slump (20–50 mm) concrete should be used. For slopes between 25° and 35° it would be advisable to reduce the slump to between 0 mm and 40 mm. This may appear a rather wide range, but it is not practical to control the slump within closer limits than about ± 20 mm.

In water reservoirs it is usual for the perimeter bays to form the foundation for the wall panels and these may then have a slope of 1 in 5. An effective method of constructing wall panels to help reduce the incidence of thermal contraction cracking is to concrete the base and the panel in one operation as in this way restraint between the base and the wall panel is reduced to a minimum. The authors have seen this done without the use of a top shutter on the surface of the sloping base slab.

As mentioned above, for slopes exceeding about 35°, a top shutter is likely to be necessary and this certainly creates some difficulty with compaction. The work has to proceed slowly up the slope and only a comparatively short length of slab can be properly compacted with poker vibrators beneath the shutter. This means that the shutter must move up with the concrete. Because of the difficulties involved in this it is advisable to strip the shutter as soon as the concrete has stiffened sufficiently to be self-supporting on the slope in question but not so stiff that necessary finishing operations cannot be carried out.

A commonplace problem when placing concrete under an inclined shutter is the entrapment of air below the shutter. Experience shows that there is no simple way of eliminating this. However, measures can be taken to reduce the possibility of entrapping air. A very workable concrete (having a slump not less than 75 mm) should be used. The concrete should be placed in a slow, steady stream away from the underside of the top shutter, as far as this is practical. The object is to obtain a maximum degree of self-compaction before flowing against the top shutter. Concrete of this type requires a comparatively high cement content and the slump should be increased by the use of a plasticizer and not by raising the water/cement ratio.

8.8. *Walls and Columns*

General Considerations

Problems associated with thermal contraction cracking and the closely connected subject of joints in walls have been discussed in some detail in Chapter 7.

It is usual for the base of the wall of a liquid retaining structure to form the perimeter bay of the floor. This base can be cast first, with the wall panel following some time later, the period varying from a few days to several months. An alternative method is to cast the base and the wall panel in one continuous operation, as mentioned in a previous section (p. 180).

When the base and wall are cast separately, the joint between the two is considered either as a monolithic joint or as a sliding joint (reference should be made to Chapter 7), depending on the basic design of the structure. It is most unusual for a tank other than one which is circular on plan, to have a sliding joint between the base and the wall.

There are generally few problems associated with the construction of columns apart from possible plastic settlement cracking at the top of mushroom headed columns (see Chapter 7), and surface finish. Columns are usually cast in one operation for their full height.

Figure 8.14 shows columns in a large service reservoir.

Reinforcement

The detailing of the reinforcement for a selected number of structures is dealt with in Part 1 of this book.

Questions often arise on site regarding the need or otherwise to remove rust from reinforcement. It is generally acknowledged that it is in the 'early

Fig. 8.14. *Large service reservoir: general view showing mushroom headed columns.* (*Consulting Engineers: John Taylor & Sons; Contractors: Costain Civil Engineering. Courtesy: Colne Valley Water Co., Rickmondsworth.*)

age' of the concrete, i.e. the first 48 h after casting, that thermal contraction cracking occurs. A matter on which information is lacking is the effect on long-term durability of initially embedding rusty steel in concrete. Treadaway and Russell in their Current Paper CP 82/68—*Inhibition of the Corrosion of Steel in Concrete*, state that the corrosion rate of steel in concrete is determined by three factors; one of these three is

> The presence of anodic and cathodic sites on the metal in contact with the electrolyte; this is a function of surface metal variations (oxide coating in contact with bare metal surface) and environment.

The authors do not think that a light coating of powdered rust will adversely affect the durability by initiating corrosion of the steel after the concrete has been cast around it. However, rust scales and pitting are a different matter, and in view of the expense and general dislocation of operation of the structure should corrosion of the reinforcement occur at a later date, it is recommended that if there is any doubt, it is better to remove the rust rather than allow it to remain on the steel.

The removal of rust from reinforcement is a time consuming operation, but with the advent of high pressure water jets and oxy-acetylene flame treatment for cleaning and de-rusting structural steel, the time factor can be drastically reduced. The authors have seen many sites where the reinforcement has been left lying around on the ground half buried in mud, then the contractors complained that the resident engineer was being unreasonable in insisting on proper cleaning of the steel.

The normal cover to reinforcement in liquid retaining structures is 40 mm on the water face and 25 mm on the other face. In the case of tanks containing sea water or aggressive liquids it is usual to specify 50-mm cover on the liquid side. This is the cover to all steel reinforcement, and not just the main bars. The authors are not in favour of cover exceeding 50 mm, because generally the closer the distribution steel is to the surface of the concrete, the better the crack control it can exercise.

Formwork

A considerable amount has been written about formwork, its fixing, and the effect of various release agents on the surface appearance of the concrete when the forms are removed. Only a few matters, which in the authors' experience are sometimes overlooked or misunderstood, will be discussed here.

If the designer wishes to reduce as far as possible, the incidence of blow holes on the surface of the concrete, then a slightly absorbent form face should be used. Steel, fibreglass and impervious sealed-faced plywood are likely to result in more blow holes than a normal ply. On the other hand the impermeable-faced forms will give a very smooth, closed surface (apart from the blow holes).

It goes without saying that the formwork must be designed to withstand the pressure of the plastic concrete. This pressure will depend on a number of factors of which the principal ones are height of pour and rate of stiffening of the concrete. The latter is of particular importance if there is any change in mix proportions and/or type of cement, or inclusion of admixtures, or dilution of the cement by the addition of PFA or granulated blast furnace slag. These latter materials have pozzolanic properties and may not reduce the long-term strength of the concrete. However, they will reduce the rate of gain of strength up to about 14 days after casting; this can increase the pressure on the formwork compared with concrete made entirely with ordinary Portland cement. Reference should be made to Chapter 5 for more details on the use of PFA in concrete.

A matter on which there is sometimes controversy relates to the use of through-ties for the fixing of the formwork. Some engineers do not allow through-ties at all and so the formwork has to be secured by other means, such as struts, walings, etc. The authors do not consider that the prohibition of through-ties is justified provided that these are the type which snap off and leave the central portion embedded in the concrete; the two end pieces and the rubber cones are removed, and with some makes these sections can be re-used. Some makes of wall tie contain a 'puddle' flange or water stop in the centre, but it is doubtful if this has any practical effect on the watertightness of the bolt hole. The efficient plugging of the hole is the key to watertightness.

Bolt holes should be cleaned out and then plugged, using either an earth-dry cement/sand mortar (1:3), or a cement/sand mortar (1:3) gauged with a styrene butadiene latex (10 litre latex to 50 kg cement). Two brush coats of latex grout should be brushed over and around each filled hole as late as possible after plugging.

To ensure correct cover to the steel, spacers should be securely fixed to the reinforcement prior to the erection of the formwork.

Casting the Walls and Columns

The authors have already commented on the concept in the Code (BS 5337) relating to degrees of exposure and corresponding minimum cement contents. Unless conditions are aggressive to the concrete, it is considered that a more logical and technically sounder approach is to relate the cement content to the wall thickness. The thickness selected for the change in cement content is empirical and there is nothing absolute about the figures suggested here.

For reinforced concrete walls up to say 600 mm thick, the minimum cement content can be 360 kg/m^3, and this can be reduced to 330 kg for

Fig. 8.15. *View of PVC water bars in horizontal joints in walls of deep sump. (Courtesy: Servicised Ltd, London.)*

thicknesses between 600 mm and 1000 mm. For greater thickness the cement content can be further reduced to 300 kg or lower. The water/cement ratio should be maintained at a maximum of 0·50. A slump of 50 mm ± 25 mm is likely to be adequate to secure proper compaction.

For the columns, the mix should contain not less than 360 kg/m^3 with a water/cement ratio of 0·50, and a slump of 50 mm ± 25 mm. Admixtures should only be permitted with the written approval of the resident engineer and in any case should be chloride-free. Air entraining admixtures can be useful in providing a cohesive mix and preventing water scour; the air content should be $4\frac{1}{2}\% \pm 1\frac{1}{2}\%$.

See also the notes on mix proportions for the floor slab on p. 219 and the section on mix design (p. 158).

Walls and columns can be cast satisfactorily in one continuous operation to heights of 8 m or even 10 m. This has the advantage of eliminating horizontal 'cold' joints in the majority of liquid retaining structures.

For these high pours, the mix should be cohesive and should not segregate during placing. Thorough compaction is most important, but it must be accepted that the concrete at the base of the wall or column will inevitably be denser than at the top of the lift. Therefore, should cores be taken at a later date, those at the bottom of the lift are likely to give higher density and strength than at the top.

Figure 8.15 shows the positioning of PVC, water bars in the horizontal joints of an exceptionally deep sump to a main sewage pumping station.

For proper compaction it is essential for the contractor to provide an adequate number of poker vibrators in good working order, with at least two in reserve in case of breakdown. One poker is required for every 1·5–2·0 m run of wall. Some engineers insist that the first few skips of concrete in a lift shall contain more cement than specified for the main part of the pour. The Code (BS 5337), Clause 14, recommends that when concreting walls, the mix proportions for the first 250-mm depth of concrete placed on a horizontal joint should contain a smaller proportion of coarse aggregate.

The concrete should be distributed evenly along the full length of each wall panel and the poker vibrators should not be used for this purpose. The concrete should be brought up evenly and if an emergency occurs before the panel is complete, the concrete should be brought to a horizontal surface. There is divergence of views on the desirability of using trunking to place concrete in deep lifts. The authors' experience is that in most cases, trunking can cause more trouble than it is worth.

Water bars must be of the type which can be securely fixed in position.

Joints

It is prudent to provide a sealing groove and seal all vertical joints in walls. In most cases, the authors feel that horizontal joints need not be sealed.

Figure 8.16 shows the preparation of the joint in the wall of a sewage tank prior to the insertion of the preformed 'Neoferma' gasket. For watertightness the sealant must bond well to the sides of the sealing groove, which means that for practical reasons, it must bond well to damp concrete.

Surface Finish

BLOW HOLES

The surface finish of the concrete as it comes from the forms is a matter which sometimes causes disputes on site. Generally, for water and sewage tanks, particularly the former, the engineer wants as smooth a finish as possible which means a surface relatively free from blow holes. It is virtually impossible to describe in a specification the exact standard of finish which will be accepted and therefore common sense is required on both sides. It is certainly necessary for the specification to be as clear as possible and it is advisable that if the standard of finish is considered to be important that this fact be emphasized in the contract documents. If this is done the contractor should take all necessary steps in consultation with the engineer to achieve the finish required. This will involve special attention to mix design, type of formwork, type of release agent, and precautions to ensure grout-tight joints in the formwork. All this preliminary investigation should be carried out as early in the contract as possible.

It should be remembered that blow holes are more likely to be troublesome when the formwork has an impermeable surface, as with steel, glassfibre, and sealed plywood, than when a more absorbent formwork lining is used.

Blow holes and surface honeycombing should be carefully filled in with a thick grout or fine mortar containing a styrene butadiene latex.

External finishes of an acceptable architectural standard for above ground concrete will be discussed under Section 9.3, p. 281.

WATER SCARS (WATER RUNS)

A condition which can cause a certain amount of consternation on site when formwork is stripped is water scars and water migration on the surface of the concrete. While this does not improve the appearance of the concrete there is no reason to believe that it is detrimental to the strength of the concrete in the wall. However, this defect may increase the risk of corrosion of the reinforcement as it is the surface layers of the concrete which are

Fig. 8.16. *Preparation of joints in wall of sewage settling tank prior to sealing with 'Neoferma' gaskets. (Courtesy: Colebrand Ltd, London.)*

affected. This type of surface defect is considered to be due to lack or absence of fines below the 52 sieve, that is, poor grading of the fine aggregate. Unless it is practical to change the source of the fine aggregate and obtain sand which is properly graded, the solution has to be arrived at by other means and by trial and error. The addition of PFA, some increase in cement content, reduction of water/cement ratio and use of a plasticizer to maintain workability, can all be useful, either individually or in some combination. The use of an air entraining admixture (see Chapter 5) has proved to be very effective in many instances.

Figure 8.17 shows a water 'scar' on a reservoir wall.

Figure 8.18 shows the kind of finish which can be achieved by a good contractor and close supervision.

Curing

It is now accepted that all structural concrete should be adequately cured, and this includes wall panels. The problem is, what is adequate curing and how should it be carried out.

As long as the formwork remains in position no special curing is required as the formwork performs this function satisfactorily. Generally, contractors are in a great hurry to remove the formwork. Even when the specification states quite clearly that the formwork must not be removed until 72 h after completion of casting, the site agent is often very indignant if he is not allowed to strip the forms within about 16–24 h of completion of the pour.

The first consideration is of course, that the concrete must be strong enough to support its own weight and withstand external stress and shock due to the removal of the formwork. Another important factor which should be given due consideration is that if the formwork is timber it will provide good thermal insulation and therefore the temperature of the concrete will be higher than if steel formwork is used. Hence, the early stripping of timber for work in cold weather can result in a rapid drop in surface temperature and a steep thermal gradient through the wall. On the other hand, in hot weather, the early removal of the formwork is likely to be beneficial by facilitating the loss of heat from the concrete. The problems associated with thermal contraction cracking have been discussed in detail in Chapter 7.

From these brief comments it will be obvious that a practical and flexible approach to the question of the most suitable time for removal of the formwork from walls is needed in the interests of a good job and good site relations.

Immediately the formwork has been removed normal curing procedures should be started. The essential factors are as follows:

(a) The concrete must be protected against rapid evaporation of moisture as this will result in a weak surface layer, possibly shrinkage cracking and the unnecessary widening and extension of any thermal contraction cracking which may have already occurred.

(b) In cold weather the concrete must be protected against thermal shock resulting from a steep thermal gradient between the centre of the wall and the surface.

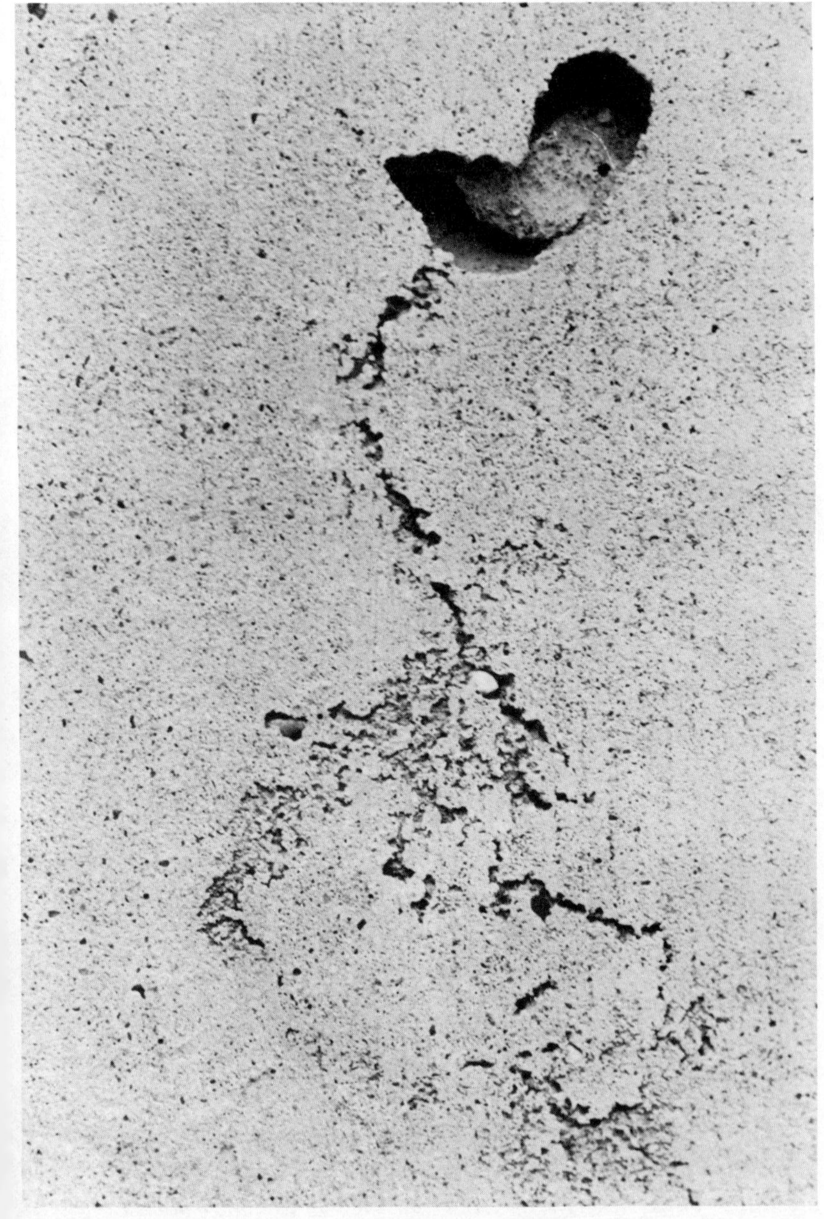

Fig. 8.17. *Example of water scouring on concrete wall.* (*Courtesy: Cement and Concrete Association, London.*)

Fig. 8.18. *Inside view of service reservoir showing excellent finish to concrete. (Consulting Engineers: John Taylor & Sons; Contractors: Costain Civil Engineering. Courtesy: Colne Valley Water Co., Rickmondsworth.)*

Requirement (a) above can be achieved by the use of a good quality resin based curing membrane. In the case of drinking water reservoirs, the membrane must be non-toxic. Special covers of 1000-gauge polyethylene sheets are effective, provided they are designed and erected to exclude the wind from blowing underneath them.

However, in cases where temperature rise in the maturing concrete has to be strictly controlled, the use of polyethylene sheets for walls should not be used due to the possible 'greenhouse' effect.

In cold weather, thermal insulating covers may be required.

8.9. *Roofs*

General Considerations

Many types of liquid retaining structures have to be roofed. These include water reservoirs of small to medium size, water towers, sludge digestion tanks and tanks holding petroleum oils and liquids used in the chemical industry. At the present time in the UK it is not the practice to roof over sewage sedimentation and humus tanks and storm over-flow tanks. However, there is increasing pressure from various groups to improve amenities and reduce smell around sewage treatment works and this is likely to result in the provision of roofs to many of these tanks.

Depending on the size and shape of the structure, the roof can be flat or sloping to a greater or lesser degree. Sloping roofs include shells, domes, and folded plates.

For structures holding drinking water it is a usual requirement that the roof shall be watertight when it is built and remain so during the life of the structure. The simplest way of ensuring watertightness is to provide the roof slab with a high quality, properly applied membrane, either *in situ* or preformed sheeting.

Even when a membrane is provided, it is recommended that reasonable precautions be taken to ensure that the structural roof slab is itself watertight. Defects resulting in leakage generally take the form of cracking caused mostly by thermal contraction in the early life of the concrete. To help prevent this it is recommended that a provisional sum be included in the contract for measures to protect the newly placed concrete from solar heat gain. This can be used at the discretion of the resident engineer and will depend on the weather at the time the roof slab is cast. The 'measures' would consist of protective covers to be put into position immediately the compaction and finishing operations are complete.

Metal Roofs

Metal roofs have been used to a limited extent, but have not proved particularly satisfactory. The conditions of exposure are severe, particularly for ferrous metal of which many of these roofs are constructed.

In the case of drinking water reservoirs, there may be a small amount of chlorine compounds present in the air above water level if the water is chlorinated before admission to the reservoir, and relative humidity is always very high. In sludge digestion tanks aggressive gases combined with high temperature can cause serious trouble in the long term, even to well protected steel.

As far as the authors are aware, there is no known method of permanently protecting steel in a very humid environment except by surrounding it with high quality Portland cement concrete. This is quite impractical for steel roof trusses. The great advantage of steel for this purpose is light weight and ease and speed of erection. Probably the most effective surface protection at the present time for water tank roofs, is to first heavily galvanize the steel by one of the approved methods and then coat the galvanized surface with PVC plastisol 0·20 mm thick, or PVC laminate 0·30 mm thick. Such a coating, if properly applied, may be expected to have a maintenance-free life of at least 15 years. *In situ* coatings of properly formulated polyurethane or epoxide resin are also very durable.

For sludge digestion tanks a nylon–polyamide 0·25–0·50 mm thick on heavily galvanized steel would give excellent protection.

The cost of these special coatings is very high.

Concrete Roofs

For the purpose of this section, reinforced gunite is considered as concrete, although in the UK it is usually a pneumatically applied mortar.

Concrete roofs can be divided into the following types:

(a) *In situ* reinforced.
(b) Precast reinforced and prestressed.
(c) A combination of (a) and (b) (precast structural units with an *in situ* topping).
(d) Reinforced gunite.

In Situ Reinforced Concrete

The standard roof for reservoirs, water towers and similar structures is reinforced concrete, either beam and slab or flat slab. Such a roof is relatively simple to design and construct, apart from the inherent difficulties

Fig. 8.19. *Large service reservoir; view of formwork for flat concrete roof. (Consulting Engineers: John Taylor & Sons; Contractors: Costain Civil Engineering. Courtesy: Colne Valley Water Co., Rickmondsworth.)*

of ensuring complete watertightness previously mentioned. The principal source of leakage in such roofs is through cracks and joints. Cracks are mostly caused by thermal contraction during the early life of the concrete, and this is often aggravated by inadequate curing.

Unless the slab is of small area it is unlikely to be cast in one continuous operation and therefore joints will be necessary. There is an increasing tendency to cast floor and roof slabs in larger and larger areas. It is now not unusual to cast 300 m^2 of high quality concrete (which means a cement-rich mix), in one operation. With a 200 mm thick slab, this would require about 60 m^3 of concrete.

Figure 8.19 shows formwork for a roof slab.

In Chapter 7, there was a detailed discussion on thermal contraction cracking. However, the thermal conditions of a suspended slab are in many ways different to those of a slab cast in contact with the ground or fill material. This difference is due principally to the insulating properties of the formwork; also the conditions of restraint are quite different.

Any full movement joints in the supporting walls should be carried through the roof slab. Apart from these, contraction or stress-relief joints may be required.

The location and detailing of these joints present a number of problems to which there is no clear-cut solution. While for slabs uniformly supported on the ground there is no technical advantage in casting the slab in alternate bays or checkerboard pattern, this does not necessarily apply to thick suspended slabs. In these cases there may be some advantage in delaying the casting of adjacent bays until the concrete has cooled and the initial thermal contraction has taken place. Generally, a period of 72 h is adequate for this. At present it is not possible to predict with any degree of accuracy the magnitude of these thermal contraction stresses, but the factors involved include the following:

(a) The cement type, and content of the mix
(b) The rate of evolution of heat of hydration of the cement.
(c) The temperature of the concrete at the time of placing.
(d) The ambient air temperature during the first 72 h after casting.
(e) The '*U*' value of the formwork, and of any covers provided on the top surface of the slab.
(f) The coefficient of expansion of the aggregates used.
(g) The amount and location of reinforcement in the slab.
(h) The method of curing the slab.
(i) The restraint on movement of the slab imposed by its supports.
(j) The area of slab cast as one monolithic unit.

In view of the difficulties involved in an accurate analysis of the magnitude and distribution of contraction stresses occurring in the slab, it is recommended that the following practical precautions be taken:

(a) Ordinary Portland cement should be used.
(b) Subject to requirements for durability and strength, the cement content of the mix should be kept to a minimum.
(c) If there is a choice of aggregates, select one with a low coefficient of expansion, such as limestone.
(d) Keep the water/cement ratio as low as possible and use a plasticizer to obtain workability if this is necessary.
(e) If the slab is cast in hot weather adopt the techniques recommended in Chapter 7.

(f) Ensure adequate curing.
(g) The whole concreting operation should be carefully preplanned. If the concrete is ready-mixed, the full co-operation of the supplier should be ensured by prior consultation and clear instructions.
(h) An adequate amount of high-grade supervision should be provided on site.
(i) Within the area of one casting unit there should be no part of the slab without adequate top reinforcement to take thermal contraction stresses. There are usually areas in such slabs where top reinforcement is not required to take stress due to dead and live loads. It is these areas which are particularly vulnerable to thermal contraction cracking. This 'special' reinforcement can be provided in the form of a fabric to BS 4483—*Steel Fabric for the Reinforcement of Concrete*. It is normally placed 40–50 mm from the top of the slab.

The designer, having considered all the factors outlined above, may still feel that due to the large area of slab involved thermal contraction cracking is likely to occur. In such a case stress-relief joints should be provided and positioned so that they will not adversely affect the structural stability of the slab under the worst conditions of loading. It is clear from this that each case has to be considered as a separate design problem. Figure 7.10 (p. 190) shows a full movement joint in a roof slab.

Regarding specification requirements for the concrete for the roof structure, i.e. the beams and slab, these can be as follows:

> Minimum cement content: 360 kg/m^3 and a maximum water/cement ratio of 0·50. The aggregates should comply with BS 882 and should be 20 mm maximum size. Admixtures, if permitted in writing by the resident engineer, should be chloride-free. The slump should be in the range 50 mm ± 25 mm.

The remarks earlier in this chapter on the need to consider actual site conditions when determining cement contents are also relevant here.

It is usual practice for the final roof covering to be laid to falls so that ponding is reduced to a minimum. The necessary gradient (a minimum of 1 in 150) can be obtained by either providing a screed or topping on the structural slab, or by constructing the roof slab itself to the required gradient. The authors favour the latter as they consider that screeds and toppings are better avoided whenever possible. The details of how this is achieved will depend on the size and shape of the roof and the position of the rainwater outlets.

Precast Concrete—Reinforced and Prestressed

The use of precast concrete elements for the roofs of liquid retaining structures has increased in recent years.

Precast units have advantages over *in situ* construction, but there are also disadvantages. The advantages include

(a) partial or complete elimination of formwork;
(b) speed of erection and absence of curing and maturing period on site;
(c) if the units are prestressed they are often of smaller section (less massive) and therefore lighter in weight;
(d) if the units are made in a factory the standard of concrete quality and finish should be higher than if they are cast in the open on a construction site.

The disadvantages include

(a) a considerable degree of standardization of dimensions is required because if the precast units are a 'once-off' job they are likely to be prohibitively expensive;
(b) adequate bearing and fixing must be provided to the supporting units such as columns and walls. These fixings must be carefully designed and detailed and may require the use of stainless steel which is expensive;
(c) some standard precast units require an *in situ* structural concrete topping when the loading or span exceeds certain limits. Propping is often required until this topping has reached its design strength. The question of possible deflexion in the precast units and the degree of rigidity obtained by propping should be given careful consideration. Assumptions made in the design may not be achieved on site.
(d) unless an *in situ* topping is used it is difficult to ensure a watertight roof due to the small movements which will occur between the numerous precast units;
(e) if they are cast away from the site, there will be transportation costs and possible damage during loading and unloading, etc.

In Situ Prestressed Concrete Roofs

The design and construction of the roofs of liquid retaining structures in this country in *in situ* prestressed concrete, whether pretensioned or post-tensioned, is very rare. The design of circular structures in prestressed concrete is not unusual, but it is only the walls that are prestressed (usually post-tensioned vertically and circumferentially). The roofs are often domed and are supported on substantial ring beams which are prestressed. The domed shape ensures that the roof slab is generally permanently in compression. This helps greatly in reducing cracking and defects are usually due to honeycombed concrete caused by segregation and/or lack of compaction.

Where roofs are flat, it seems to the authors that a post-tensioned design has many advantages from the point of view of reducing or eliminating cracking due to thermal contraction and drying shrinkage. However, it appears that this type of design does not appeal to the designers of such structures. The nearest approach to this was the roofs of two car parks in Reading designed by J. Brobowski and Partners. A paper on the prestressing of edge members in long buildings was presented to a Symposium held in Madrid in 1970; details of the paper are given in the bibliography at the end of this chapter.

In Situ Concrete Toppings to Precast Roof Units

In situ concrete toppings to precast roof units can be either structural or non-structural. In the former case the topping is bonded to the precast unit below, but in the latter case the special precautions required to ensure bond are often omitted, and in some roofs the topping is debonded from the structural slab by the interposition of a membrane.

Structural Toppings

General considerations. In precast concrete roof units the concrete is normally high quality, and the sections are designed to be as slender and light in weight as possible. This applies particularly to prestressed units. Therefore when structural toppings are used special attention must be paid during manufacture to ensure that the top surface of the units have a good key to promote bond with the topping which will be laid on them. In most roof slabs it is unlikely that the shear at the interface will be high enough to justify shear connectors. In this connection, reference should be made to CP 116—*The Structural Use of Precast Concrete*, Clauses 341 to 345. Clause 341 states that the precast units and the *in situ* topping can only be considered as acting in conjunction when provision is made for the horizontal shear at the interface by means of suitable roughening of the precast units or the provision of shear connectors. The method of preparing the surface of the precast units is detailed in Clause 409 (d), and information on calculations for shear at the interface and permitted shear stress when shear connectors are and are not provided is given in Clause 345.

Preparation of the Surface of the Precast Units

When a structural topping is added to an *in situ* concrete slab, the surface is normally scabbled, but scabbling is generally considered unsuitable for precast units due to their slim dimensions and high strength of the concrete. The newly developed technique of high pressure water jetting and flame texturing could be used if the roughening in the factory is considered inadequate by the resident engineer.

In the factory, brushing with a water spray when the initial set has taken place is often adopted but can result in a weak porous layer on the top of the unit.

If shear connectors are provided in the design then they must be cast into the units when they are made.

Other site methods for improving bond include thorough wire brushing to remove all weak laitence. The use of an epoxy resin-based bonding agent may also help, but it is essential that the topping should be placed and compacted before the bonding agent has set; the bonding layer must be still tacky when the topping is laid on it, otherwise it will become an effective debonding layer. Units which were originally properly prepared but have subsequently become dirty or contaminated must be cleaned with an industrial floor cleaner such as 'Glocrete' or 'Basol 77', or with high velocity water jets, or with flame texturing equipment.

Some types of precast units are provided with vertical keys and the concrete is cast into these simultaneously with the laying of the topping.

Mix Proportions for the Structural Topping

The concrete mix for the structural topping is normally a 'designed' one, but the following outlines what the authors consider should be the minimum requirements to ensure watertightness and durability:

Cement: ordinary Portland or rapid hardening Portland cement.

Aggregates: In most cases, the aggregates will be 'from natural sources' and therefore should comply with BS 882. However, if thermal insulation is required then a suitable lightweight aggregate could be used, probably with a natural sand to provide the necessary strength. For information on lightweight aggregates, refer to Chapter 5. For toppings which are 75 mm thick and less, a maximum size of aggregate of 10 mm can be used, but for thicker toppings a 20-mm maximum size is preferred.

Slump: 25–75 mm (i.e. 50 mm $\pm$ 25 mm).

Strength: 30 N/mm^2, subject to the minimum cement content of the mix given below, and the slump.

Minimum cement content: 330 kg/m^3 of compacted concrete; if 10-mm aggregate is used, the cement content should be increased to 360 kg/m^3.

Bay Layout and Provision of Reinforcement for Bonded Toppings

Opinions differ on the minimum thickness of bonded toppings, the bay size and the need or otherwise to provide reinforcement. It is very difficult to be precise about these matters, as so much depends on the quality of the bond with the precast concrete units as well as other factors such as amount of water in the mix, type of cement, and thickness of the topping. Some manufacturers of precast units recommend toppings as thin as 30 mm. However, the authors consider that such a thin topping is unlikely to be satisfactory for the roof of a liquid retaining structure, and recommend a minimum thickness of 75 mm. With a thickness of less than 100 mm, it is better to omit reinforcement and limit the bay size to about 4 m × 4 m, i.e. 16 m^2; the shape can be rectangular provided the length does not exceed 1·5 times the width. With a thickness of 100 mm, the bays could be a maximum 5 m × 5 m (25 m^2).

The reason for restricting the bay dimensions is that the larger the bay, the greater is the tendency for cracking, and curling along the edges and at the corners of the bays.

Where it is decided to lay a thicker topping of say 125 mm or 150 mm, the size of the bays can be increased by the provision of fabric reinforcement.

Except in 'hogging' moment areas there are no recognized rules for calculating the weight of reinforcement for these structural bonded toppings. The reinforcement is mainly required to take the stresses caused by thermal and drying shrinkage cracking, but also must resist 'hogging' (negative bending) moments where the topping crosses a support.

The use of thicker toppings and the inclusion of reinforcement allows much greater freedom in the shape and dimensions of the bays. For square bays a square mesh should be used, while for rectangular bays a rectangular mesh with the main wires parallel to the long side of the bay should be used.

For roof and floor slabs a minimum cover of 40 mm is recommended. If the topping is covered with a flexible waterproof membrane (as is often the case with roofs), then the cover can be reduced to 25 mm.

Table 8.1 indicates the minimum weight, etc., of fabric reinforcement which the authors feel should be used for bonded toppings exceeding 75 mm thick.

All full movement joints in the precast structural system of units must be carried through the topping. Where the topping is unreinforced it is advisable to either form a bay joint along the centre-line of the supporting beams and walls or to provide reinforcement in the topping to take the negative bending across the supports. For toppings less than 75 mm thick, it is better to omit the reinforcement and form a plain butt joint.

The topping must be well compacted in order to provide strength, good bond and watertightness. The method of compaction should be suitable for the thickness and size of the bays. It is not practical to use poker vibrators to compact toppings less than 100 mm thick. Therefore for these thinner toppings, hand-operated tampers working off the side forms should be used. Better results are obtained if the tamper is fitted with a vibrator. The use of tampers limits the bay width to about 4 m.

Table 8.1

Bay size		*Weight of*	*Cross-sectional area of*	*Cross-sectional area of*	*BS 4483 reference*
Length, m	*Width, m*	*fabric, kg/m²*	*longitudinal wires, mm²/m*	*transverse wires, mm²/m*	*number*
10	4	3·73	283	193	B. 283
15	4	4·53	385	193	B. 385
20	4	5·93	503	252	B. 503
20	12	5·93	503	252	B. 503
10	10	3·95	252	252	A. 252

The finish given to the topping will depend on whether it is intended to lay on it a preformed sheet membrane or an *in situ* membrane, as the waterproofing layer. A smoother finish is required when the membrane is sheet material. Methods of finishing floor slabs have been discussed earlier in this chapter and the recommendations given are generally applicable here.

After finishing, the topping must be cured by one of the methods described for floor slabs, but if it is proposed to use a spray or brush applied curing membrane, the suppliers of the final waterproof membrane should be consulted, as the presence of the remains of the membrane material may have an adverse effect on the adhesion of the waterproof membrane.

Joints in the Topping

Full movement joints should be detailed as previously described in this chapter.

The authors feel that the transverse joints between the larger reinforced bays should be provided with a sealing groove and sealed with a flexible sealant. However, for the longitudinal joints, alternate joints only need be provided with a flexible sealant. The intermediate joints can be grouted in with a cement grout composed of 50 kg ordinary Portland cement to 25 litre styrene butadiene based latex emulsion. The grouting-in should be carried out as late as possible after the completion of the topping. Before the grout is applied, the joint should be well brushed with a wire brush to remove loose and friable material.

The bay joints, both longitudinal and transverse, can be plain butt joints.

The grouting-in of the transverse joints means that a virtually monolithic area is formed, having a length equal to the bay and a width equal to that of two bays.

Waterproofing of Concrete Roofs

It is a general requirement that roofs should be watertight. Clause 33 of the Code (BS 5337) specifically recommends that roofs of structures holding potable water should be watertight and should be tested.

There are various ways of achieving this:

(a) The roof may be designed on conventional lines, using the elastic theory and in accordance with the principles set out in Clause 9 of BS 5337, and then to rely on the watertightness of the concrete.

(b) The designer may design on 'limit state' principles, as set out in Clause 8 of BS 5337, which allow cracks of 'calculated' width and spacing, and trust that the theory is correct and that no leakage will occur.

(c) The roof may be designed to either (a) or (b) above, and then provided with a watertight membrane.

The authors favour the alternative in (c) as this is the standard to which roofs of buildings are designed and these have to be completely watertight. It may be argued that if the roof of a sewage or sludge tank leaks slightly it does not really matter as no harm will come to the contents. However, the effect of even slight seepage over a long period on the steel reinforcement must be given careful consideration.

There is now a wide range of materials available for waterproofing concrete roofs. These include sheet materials such as built-up roofing felt (CP 144: Part 1), synthetic rubber of many types, butyl sheeting, poly-isobutylene, chlorinated polyethylene, and PVC; the *in situ* compounds include mastic asphalt (CP 144: Part 2), and polymer resins such as epoxides and polyurethanes.

When used in the appropriate circumstances, and correctly laid, all these materials will give good service.

In selecting the membrane, the following factors should be considered:

(a) Whether the membrane will be exposed or will be covered (and thus protected from the weather and sunlight) by another material.
(b) Whether there are any special requirements regarding wear/abrasion.
(c) The degree of flexibility required in the membrane.
(d) The shape, gradients and accessibility of the roof for the original laying of the membrane and for any subsequent repairs.

Sheeting can be laid fully bonded to the base concrete, partially bonded or unbonded. The recommendations of the suppliers should be followed. The authors recommend that a material should be selected which can be fully bonded to the concrete with a non-moisture-sensitive adhesive. The advantage of full bond is that, should the material become damaged, leakage will not occur unless there is a defect in the concrete immediately below the break in the membrane, a most unlikely event. The advantage of semi-bonded or unbonded membranes is that movement in the concrete slab does not induce stress in the membrane.

Mastic asphalt is normally laid on a separating layer and so from a stress point of view is similar to an unbonded sheeting. The usual thickness is 20 mm and this weighs about 42 kg/m^2. Bitumen-felts can be partially or fully bonded to the concrete, and weigh between 2·0 and 3·0 kg/m^2. Such materials as butyl rubber and polyisobutylene are usually about 0·75 mm thick, and the same applies to synthetic rubbers like Hypalon. These are usually fully bonded to the concrete deck, and the sheets are solvent welded.

Polyvinylchloride (PVC) and chlorinated polyethylene have only come onto the roofing market in the UK in recent years although they have been used extensively on the Continent and elsewhere. The sheeting is usually laid unbonded except at the perimeter, although sometimes it is spot-bonded to the structural slab. This type of sheeting is often used in the 'upside down' method of roof construction in which the thermal insulation is laid above the membrane. In this arrangement the thermal insulation must be impervious to moisture.

Sheeting can be laid quickly and in very cold weather, and in the right conditions will give good service.

However, the *in situ* polymer resins, epoxide and polyurethane, are much tougher and in spite of their relatively high cost are to be preferred for civil engineering structures.

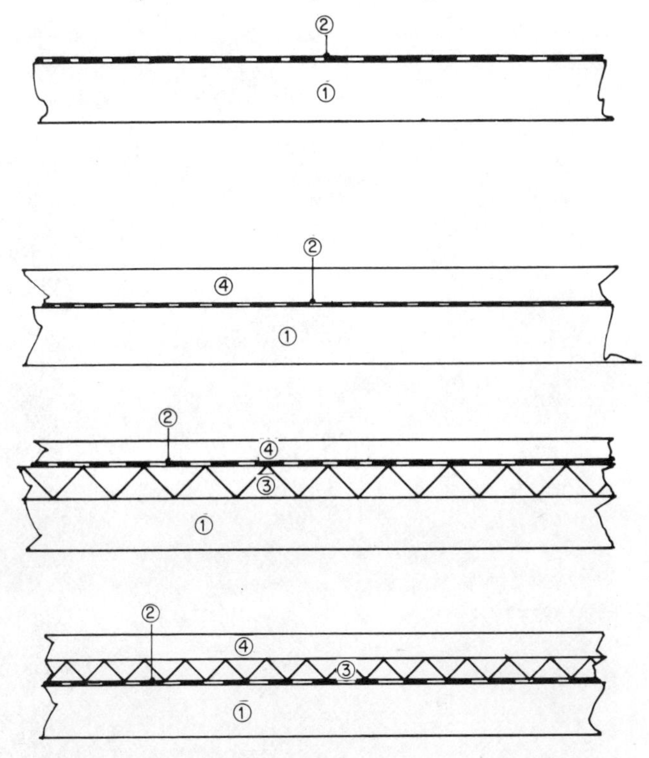

Fig. 8.20. *Methods of finishing flat concrete roofs.* 1, *Structural slab laid to falls;* 2, *waterproof membrane;* 3, *thermal insulation (moisture-resistant);* 4, *wearing surface.*

For all bonded toppings, the surface of the base concrete must be properly prepared by the removal of all weak laitence, dirt and other contamination. They can be applied by brush or airless spray, and the usual thickness is in the range 0·25–0·75 mm in three coats.

Regarding the use of cement/sand screeds on flat roofs, it is recommended that these be omitted whenever possible, as they are often a source of trouble. The necessary falls to allow rain-water to run off can be provided in the structural slab. The minimum gradient is about 1 in 150.

Figure 8.20 shows some alternative methods for finishing flat concrete roofs.

The laying of an *in situ* polymer resin membrane to the roof of a reservoir is shown in Fig. 8.21.

The type of topping provided to protect the membrane will be largely

Fig. 8.21. *Application of polymer resin membrane to concrete roof of reservoir. (Courtesy: Tretol Building Products Ltd, London.)*

determined by the use to which the roof will be put. In some cases gravel or earth is placed directly on the membrane. Precast hydraulically pressed concrete paving slabs can be used, and also *in situ* concrete. Special care in the selection of the topping is required if the roof will be used for vehicular traffic, as it must be abrasion resistant and stand up to de-icing salts.

A question arises with regard to the waterproofing of steeply sloping roofs such as shells and domes, folded plate roofs and similar. Water will run off very quickly and if after a careful inspection no defects are found, is a membrane really necessary? As previously stated, the Code (BS 5337), Clause 33, requires the testing of roof with water in cases where the structure will hold potable water. The same clause recommends the provision of a membrane when a water test cannot be carried out, for example domed roofs. The authors wonder whether the provision of a membrane is really required for domed roofs and suggest that a thick brush coat of grout made with Portland cement and a styrene butadiene latex emulsion (in proportion of 50 kg cement to 25 litre of the emulsion) will give adequate protection against infiltration of rain water.

Bibliography

American Society for Testing and Materials. *Standard Specification for Liquid Membrane-forming Compounds for Curing Concrete*, C 309–73.

Bobrowski, J. and Abeles, P. W. Prestressing of edge members in long buildings. *Symposium of International Association for Bridge and Structural Engineering Design of Concrete Structures, Madrid*, September 1970, pp. 201–208.

British Standards Institution. *Commentary on Corrosion at Bi-metallic Contacts and its Alleviation.* PD 6484, London, 1979, 26 pp.

British Steel Corporation. *Building with Steel*, No. 9, February 1972. Theme—Anti-corrosion, pp. 2–22.

Campbell, R. H., Harding, W., Misenheimer, E. and Nicholson, L. P. Surface popouts: how are they affected by job conditions. *J. ACI*, June 1974, pp. 284–288.

Concrete Society. *Falsework*. Report of the Joint Committee of the Concrete Society and the Institute of Structural Engineers, London, 1971, 52 pp.

Concrete Society. *Concrete Pressure Graph for Formwork Design*, 1972, 3 pp.

Department of Transport. *Specification for Road and Bridge Works*, Fifth Edition. HMSO, London, 1976.

Harrison, T. A. *Tables of Minimum Striking Times for Soffit and Vertical Formwork*. CIRIA Report No. 67, October 1977, 24 pp.

Harrison, T. A. *Formwork Striking Times—Methods of Assessment*. CIRIA Report No. 73, October 1977, 38 pp.

Higgs, J. B. and Hollington, M. R. *Designed and Detailed.* Code 43.501, Cement and Concrete Association, London, 1973, 28 pp.

Lilley, A. A. *Cement Stabilized Materials*. Current Practice Sheet 19, Reprint No. 5/74, Cement and Concrete Association, London, 1 p.

Murphy, F. G. *The Effect of Initial Rusting on the Bond Performance of Reinforcement*. CIRIA Report No. 71, November 1977, 28 pp.

Perkins, P. H. *Floors—Construction and Finishes*. Cement and Concrete Association, London, 1973, 132 pp.

Perkins, P. H. *Concrete Structures: Repair, Waterproofing and Protection*. Applied Science Publishers, London, 1977, 302 pp.

Perkins, P. H. *Swimming Pools*, Second Edition. Applied Science Publishers, London, 1978, 398 pp.

Russell, P. *Curing Concrete*. Code 47.020, Cement and Concrete Association, London, 1976, 8 pp.

Terzaghi, K. and Peck, R. B. *Soil Mechanics in Engineering Practice*, Third Edition. John Wiley & Sons, New York, and Chapman and Hall, London, 1968, 752 pp.

Tomlinson, M. *Foundation Design and Construction*, Third Edition. Pitman Publishing, London, 1975, 785 pp.

Treadaway, K. W. J. and Russell, A. D. *Inhibition of the Corrosion of Steel in Concrete*. Current Paper CP 82/68, Building Research Station, December 1968, 5 pp.

Weaver, J. and Sandgrove, B. M. *Striking Times of Formwork—Tables of Curing Periods to Achieve Given Strength*. CIRIA Report No. 36, October 1971, 76 pp.

West, A. S. *Piling Practice*. Butterworth, London, 1972, 114 pp.

CHAPTER 9

PRESTRESSED CONCRETE TANKS, REINFORCED GUNITE TANKS AND WATER TOWERS

9.1. *Prestressed Concrete Tanks*

Introduction

Prestressed concrete was a natural development of reinforced concrete. It was not until the late 1920s that engineers started serious work on overcoming the main inherent weakness of concrete as a structural material, namely its low tensile strength. It was obvious that this could be achieved in theory by putting the concrete into compression in such a way that some residual compression would remain even when the section was carrying its maximum design load. This technique became known as prestressing the concrete, and the resultant material as prestressed concrete.

Real progress in prestressed concrete followed the work of Freyssinet and Magnel in the period from about 1930 onwards. As with most developments in the building and construction industry, progress was slow and the exact beginnings of the new technique could not be clearly established. It is thought that the first prestressed concrete tanks were constructed in the United States in the late 1920s by a W. S. Hewitt who used high tensile bands with a turn buckle connection to tighten them around the outside of the circular concrete walls.

Although prestressed concrete pressure pipes are not considered as liquid retaining structures for the purpose of this book, it is interesting to refer to a meeting which was held by the Institution of Mechanical Engineers in 1936. This meeting was reported in the April 1936 issue of *Concrete and Constructional Engineering*, from which the following is an extract:

> Before the paper was read, Mr R. H. Harry Stanger tested a pipe made by M. Freyssinet's process. The pipe was 17 in internal diameter and $1\frac{3}{8}$ in thick

> and reinforced with bars equivalent in sectional area to a plate $\frac{1}{16}$ in thick. After vibration, the concrete of the pipe tested was subjected to a pressure inducing in the reinforcement a permanent tension of 170 000 to 185 000 lb/in^2 counteracted by a permanent compression of the concrete amounting to about 7000 lb/in^2. Mr Stanger said that a normal pipe of these dimensions would be expected to fail at about 125 lb/in^2 internal pressure, but when M. Freyssinet's pipe was tested it withstood about 1250 lb/in^2 before seepage occurred.

One of the major problems in the early prestressed concrete was the high loss of prestress in the steel due to creep in the concrete. High tensile wire for prestressing was introduced in the late 1930s and this was accompanied by a general improvement in the quality and strength of concrete.

In Scandinavia a number of *in situ* prestressed concrete swimming pools have been built, and in Germany and Switzerland precast concrete sections for floor and walls, post-tensioned together, have been used for public swimming pools. Figure 9.1 shows such a pool under construction.

General Considerations

Prestressed concrete liquid retaining structures are usually circular on plan.

The most widely used type of prestressed concrete tank in the UK is designed and constructed in accordance with the Preload system. This is essentially a circular tank with the walls circumferentially prestressed. A more detailed description of this technique, some parts of which are covered by patents, is given later in this chapter.

Apart from the Preload tanks, prestressed concrete liquid retaining structures are all 'once-off' projects, each one being specially designed. An example of design is given in Part 1 of this book.

Durability

The authors have found it surprising that as a circular shape combined with prestressing is ideal structurally for liquid retaining structures, comparatively few are built in this way. From discussions with designers and contractors, there appears to be some doubt about the long-term durability of the circumferential prestressing wires. These wires are wound around the outside of the walls, either as a continuous operation or in separate recessed bands. Protection of the wires in both cases is normally provided by gunite (pneumatically applied mortar). This guniting operation is vital for the durability of the wires and consequently for the structural stability of the tank. It is obvious, therefore, that the work should only be entrusted to an experienced firm and that site supervision must be directed to ensuring that

Fig. 9.1. *Precast post-tensioned swimming pool under construction.* (*Courtesy: Dyckerhoff & Widmann, München.*)

Fig. 9.2. *Application of cement mortar cover coat to prestressing wire on banded tank. (Courtesy: Preload Ltd, London.)*

Fig. 9.3. *View of prestressed concrete industrial tank showing banded method for locating prestressing wires. (Courtesy: Preload Ltd, London.)*

the wires are fully embedded in the gunite and all necessary precautions taken to ensure protection of the steel. If this is done properly then experience shows that durability is similar to that of reinforced concrete. Figure 9.2 shows cement mortar being applied as a cover coat to prestressing wires in a 'banded' tank. Figures 9.3 and 9.4 show completed prestressed 'banded' tanks.

The minimum distance between successive lines of wire strand is 6 mm ($\frac{1}{4}$ in). The British Code, BS 5337, Clause 4.10, requires a minimum cover of 40 mm, although in the US 25 mm is considered adequate. There are two methods of circumferential prestressing, namely

(a) continuous winding, and
(b) banding.

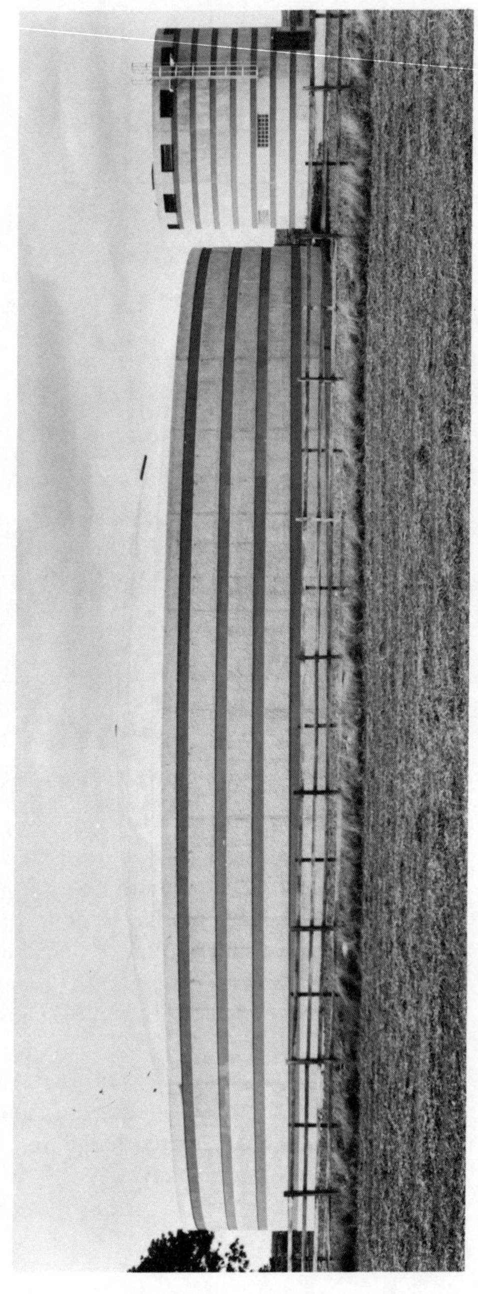

Fig. 9.4. *View of large diameter prestressed reservoir. (Courtesy: Preload Ltd, London.)*

In continuous winding, there is usually one layer of wire or strand, although there may be two. With banded tanks it is quite usual for two or more windings to be used; this method has now largely replaced the continuous winding technique. In this case a 'flash' coat of gunite is applied before the first winding, and the wire or strand is wound on to this before the gunite has hardened. On each subsequent layer, a flash coat is also applied but is allowed to harden for a minimum period of 12 h before the next layer is wound. The final protective coat of gunite is only applied when the tank is full and the walls are fully extended by the pressure of water. Figure 9.5 shows prestressing wire being wound on to a banded tank.

The usual mix for gunite is 1 part of ordinary Portland cement to 3 parts clean, sharp, well graded sand by weight, with a water/cement ratio not exceeding 0·40. British Standard 5337, Clause 28.4, requires that the mortar cover coat should be adequately cured for 7 days. Sometimes the gunite is lightly worked up with a wooden float. As soon as possible after the completion of the finishing operation, the gunite should be covered with polyethylene sheets. The sheets must be fixed so that the wind does not blow underneath them. If the weather is hot and sunny, sun covers should be provided; protection must also be provided against rain. It is unfortunate that some contractors tend to neglect these necessary precautions.

When corrosion of the prestressing wires does occur, it is usually found to be due to either the presence of chlorides in the sand used for the mortar cover coat, or to some form of incipient cracking of the mortar (probably the result of inadequate curing) or inadequate cover. The fine hair cracks may not be noticed at the time the tank was commissioned, and then due to weathering, the cracks over the years widen, admitting water which reduces the passivation provided by the cement paste, and this allows corrosion of the prestressing wires to start.

Some authorities consider that the presence of fine hair cracks in the cover coat to the prestressing wires will not result in corrosion, even in the long term. However, the authors recommend that all cracks which are visible to the naked eye should be properly sealed, either with a grout consisting of styrene butadiene latex and Portland cement (1:2 by weight) or by a slightly flexible, low viscosity, epoxide or polyurethane resin. The use of a high quality durable sealing coat over the mortar, provided it possesses the necessary degree of flexibility, will give additional protection.

There is a choice of coatings on the market, mostly based on bituminous emulsion and organic polymers such as polyurethanes. For maximum efficiency, the coating should be sufficiently elastic to endure without rupture the cycles of movement in the wall as the tank fills and empties. It is

Fig. 9.5. *Prestressing wires being wound onto banded tank* (*Courtesy: Preload Ltd, London.*)

therefore important that when such coatings are specified the maximum expected extension between empty in the cold weather and full in hot weather should be clearly stated so that the suppliers will know what is required of the coating (Fig. 9.6 refers). The use of a low viscosity resin to ensure penetration into the mortar, is also desirable.

Of equal importance for durability is complete watertightness of the wall against the passage of the liquid contained in the tank. Therefore the check for signs of leakage and seepage during the water test should be particularly stringent. It should be remembered that although the prestress will tend to close vertical cracks, it will not help if there are horizontal cracks, nor will it seal honeycombed or under-compacted concrete. All areas of seepage, sweating, etc., must be completely sealed off on the inside of the tank before it is accepted as satisfactory. This should be written into the contract documents.

Regarding the actual application of the gunite, practice in this country is different to that in the United States. In the US gunite is known as 'shotcrete', and there is a Code for this, namely American Concrete Institute Code ACI 506-66—*Recommended Practice for Shotcreting*. There is no British Code for guniting, but the Association of Gunite Contractors (see Appendix 3) has produced a specification for gunite, and the Concrete Society is preparing a code. In the United States with continuously wound tanks it is the practice to apply the shotcrete in a number of coats. The first is a 'flash' coat which is sufficiently thick to provide a cover of about 3 mm over the prestressing wires. This is followed by a 'body' coat to a minimum thickness of 12 mm and on this is applied the 'finish' coat which is not less than 8 mm thick. The total thickness of the cover to the wire should not be less than 25 mm. For more detailed information on United States practice, the reader is referred to American Concrete Institute Report No. 67–40—*Design and Construction of Circular Prestressed Concrete Structures*, and to the bibliography at the end of this chapter.

It is difficult to see the need for the application of the shotcrete in three coats. The reason for the first 'flash' coat is to provide early protection to the pressing wires after completion of winding. There seems no reason why the final cover of 25 mm should not be applied in one coat as soon as the 'flash' coat has matured, say 48 h.

The authors feel that there is some merit in the use of galvanized prestressing wire for this type of structure, as this ensures that corrosion will not start even if the application of the cover coat is somewhat delayed.

The cover coat is best applied when the tank is full and this is always done in the Preload system which also uses galvanized prestressing wire.

Fig. 9.6. *Close-up view of textured and decorative finishes on prestressed reservoir (Courtesy: Preload Ltd, London.)*

However, the use of galvanized wire should not be accepted as a reason to omit or reduce the other precautions set out in this section.

When the walls are prestressed vertically, the ducts must be grouted in. The grouting should be carried out from the lower end of the duct. Before grouting is commenced the ducts must be checked to ensure that they are clean and free of obstruction, and this includes ice in very cold weather. In view of the importance of ensuring the long-term durability of the prestressing cables, spot checks should be carried out, after grouting has been completed, by gamma radiography.

The floors and walls of prestressed tanks are almost invariably constructed of *in situ* reinforced concrete, designed and cast as for reinforced concrete tanks. A sliding joint is normally provided between the base of the wall and the foundation. The wall may be supported on the perimeter bays of the floor slab or on an independent foundation. The latter is often adopted for large tanks.

For detailed recommendations on floors for reinforced concrete tanks, reference should be made to Chapters 7 and 8.

The Walls

IN SITU CONCRETE

The walls are constructed of *in situ* reinforced concrete in this country. In the United States, the walls are sometimes constructed in reinforced gunite.

The reinforcement is 'nominal', that is, it is designed to ensure stability of the wall during construction, and the prestressing is provided to carry the operational loads. In some designs, the walls are vertically prestressed as well as circumferentially.

The sliding joint at the base of the wall must be carefully detailed and reference should be made to Section 7.5 on joints. The importance of reducing frictional resistance at this point cannot be over-emphasized; watertightness is also essential.

In the Preload system, a centrally located rubber water bar is used and the wall sits on a special sponge rubber pad.

A typical detail for the junction of the wall and floor is shown in Fig. 7.6 (p. 184). If a full movement or sliding joint is provided at the top of the wall, which is the more usual practice, then Chapter 7 should also be consulted. In Preload tanks, the roof is domed and is often carried on corbels inside the wall.

The decision as to whether to carry the roof on corbels or to support it directly on the top of the wall depends largely on the diameter of the tank. Up to 45·0-m diameter, corbels are generally used. A prestressing band is provided on the outside in line with the resultant thrust of the dome roof supported on the interior corbels, while for larger tanks the roof has a dome ring which is itself prestressed independently of the wall.

The walls can be concreted in one lift up to a height of about 8 m. A problem, however, arises as to the length of the wall panels and how joints between the panels should be detailed. In Chapters 7 and 8 there were detailed discussions on the subject of thermal contraction cracking in vertical walls and recommendations were made as to how to reduce the incidence of such cracking. This was all related to normal reinforced concrete, particularly where the main steel was vertical and the distribution steel horizontal, and the wall was fully bonded to the base.

The question now arises, if in a circular circumferentially prestressed wall, the wall cracks vertically in its early life before prestressing, does this really matter? The wall will be put into permanent compression and the cracks will be closed by this compression. In addition, under normal operating conditions for water and sewage tanks, a certain amount of 'reverse drying shrinkage' (which may also be termed reversed moisture movement) will take place, which will also tend to close such cracks. The author's opinion is that the occasional fine vertical crack in a wall panel which is later post-tensioned, is not important, but it should be dealt with as described below. Persistent and serious vertical cracking is a different matter and should be prevented by taking the measures set out in Chapter 7. If, however, in accordance with usual practice, the wall is 'free' at the base, then the chances of this thermal contraction cracking occurring is greatly reduced.

It is worth while recording in this connection that Preload tanks are often constructed with wall panels 6–10 m in length, and that cracking seldom occurs; the panels often have no more than about 0·3 % distribution steel.

Any vertical cracks which do occur should be dealt with in the following way:

(a) Thoroughly wire brush the surface of the concrete on both sides of the wall for a distance of 300 mm on each side of the crack.
(b) Lightly run a cold chisel down the crack in case there are any voids beneath the surface skin.
(c) Thoroughly clean the wire brushed surface and crack with compressed air to remove all dust, etc.
(d) Apply a coat of low viscosity epoxide resin on the prepared areas and allow to mature (cure).
(e) Apply a thick brush coat of high-build epoxide resin to both faces of the wall over the prepared area.

This work should be carried out immediately prior to the application of the circumferential prestressing.

The resin used should be non-toxic and non-tainting if the tank holds potable water.

Precast Concrete

Some circular prestressed concrete tanks have been constructed with precast wall panels. These panels can be precast on site or in a precast works.

The panels are erected by crane or gantry with short gaps between adjacent panels. The panels are supported by temporary timbering and formwork erected between the panels so that the gaps can be filled with *in situ* reinforced concrete. Reinforcement projects from the precast panels into the *in situ* concrete. The length of the *in situ* panels would vary from about 300 mm to 1·500 m. The authors are not in favour of gaps shorter than 1·00 m because of difficulty in casting the concrete, as the height of the panels may be between 6 m and 10 m.

There would be obvious advantages in time and expense if the precast panels abutted each other and the *in situ* panels were eliminated. There are, however, practical difficulties in the way of this. The principal objection is that even if a rubber gasket were used between the panels the compression induced by the prestress could cause spalling unless the adjacent (contact) faces were identical. While no doubt panels could be cast in this way, the extra cost would make serious inroads into the anticipated saving created by the omission of the *in situ* concrete.

Gunite

In the US prestressed circular tanks are sometimes constructed of reinforced shotcrete (gunite). In such cases it is usual for the shotcrete to be 'gunned' against a thin steel diaphragm. The ACI Report (67–40) recommends that the shotcrete should be built up in layers of about 50-mm thickness. The authors do not know of any tanks of this type having been constructed in the UK.

General information on gunite construction is given later in this chapter.

The Roof

The roofs of circular prestressed concrete tanks are usually dome-shaped and are designed as relatively thin shells constructed of *in situ* reinforced concrete or more rarely reinforced gunite.

In the case of heated sludge digestion tanks, thermal insulation may be

Fig. 9.7. *View of prestressed bund wall to ammonia tank* (*Courtesy: Preload Ltd, London.*)

required in the roof slab and consideration can then be given to the use of structural lightweight aggregate concrete instead of providing thermal insulation as a separate layer on the structural slab.

In a dome, the concrete is in compression, and therefore some engineers consider that provided the concrete is high quality, there is no need for a waterproof membrane even for tanks holding potable water. The quality of the concrete should be as laid down in the Code, BS 5337, Clause 4.9 and Table 8. Clause 33 of BS 5337 requires that domed roofs of structures holding potable water should be provided with a membrane.

In Chapter 8 there are detailed recommendations for the waterproofing of concrete roofs and these are also applicable to prestressed concrete tanks.

Some structures of the type considered here are located in very exposed positions. The use of air entrained concrete can therefore be considered, as this is effective in preventing damage by frost. The dosage of the admixture must be strictly controlled so that the air content is in the range of $4\frac{1}{2}\% \pm 1\frac{1}{2}\%$. The use of air entraining admixtures will improve workability

Fig. 9.8. *Aerial view of prestressed bund walls to tanks holding industrial liquids. (Courtesy: Preload Ltd, London.)*

with a constant water/cement ratio, but there will be a small reduction in the compressive strength, and a reduction in permeability.

Bund Walls

These are strictly speaking not liquid retaining structures as they are only intended to act as a second line of defence should a serious defect appear in the main tank. They are normally built around tanks which hold dangerous or inflammable liquids, particularly when the tanks are constructed of steel. There is a section later in this chapter on concrete tanks for holding petroleum oil.

Figures 9.7 and 9.8 show prestressed concrete bund walls.

9.2. *Reinforced Gunite (Sprayed Concrete)*

General Considerations

Gunite sprayed concrete, or shotcrete as it is known in the United States, is a mortar or concrete conveyed through a delivery hose and projected at high velocity by compressed air on to a surface, the force of the impact compacting the material. The principal structures constructed in gunite are shell roofs, swimming pools, tunnel linings, and irrigation canals and channels, although a limited number of water tanks and other liquid retaining tanks have been built in the UK in this material. Gunite is also used extensively for the repairs and strengthening of all types of concrete structures.

The compressive strength, usually measured from cores, should not be less than 30 N/mm^2, but may be as high as 55 N/mm^2 at 28 days.

From published information it appears that gunite is used much more extensively in the United States, Australia and South Africa than it is in the UK.

There are a number of advantages in using reinforced gunite as compared with reinforced concrete, and also some disadvantages. These may be summarized as follows:

Advantages

(a) Very little, if any, formwork is needed; this is particularly important in view of the high price of timber.
(b) The structure can be free-formed, which is important with certain types of private and public swimming pools.
(c) The structure can be built on a congested site, or one with very limited access, as the delivery hose to the 'gun' can be at least 100 m long.
(d) The speed of construction is high. A tank 30 m × 15 m × 5 m deep can be 'gunned' or 'shot' in a prepared excavation in about 20 working days.
(e) The number of joints which require detailing and sealing are reduced compared with a similar structure built in *in situ* reinforced concrete.

Disadvantages

(a) In the UK the number of experienced gunite contractors is limited, and the work should only be entrusted to such contractors.

(b) There is no Code of Practice for reinforced gunite in the UK and so the basis of design is decided by the designers themselves. This is usually the specialist contractor who always offers a design service. The Association of Gunite Contractors has produced a specification, and the Concrete Society is drafting a code which may be published in 1979. There is an American Code for Shotcrete, prepared by the American Concrete Institute—*Recommended Practice for Shotcreting*, No. ACI 506–66.

(c) In the UK the maximum depth of liquid retaining structures so far constructed in gunite is about 4·0 m. This makes the construction of large capacity tanks uneconomic in the use of ground.

(d) The variation in compressive strength and in the mix proportions of the in-place gunite is likely to be greater than in good quality *in situ* concrete.

(e) The bulk density of gunite (weight per unit volume) is lower than in concrete made with natural dense aggregates, by about 14%. Therefore, when there is a high water table, the danger of flotation is increased. This can be overcome by the installation of pressure-relief valves, but only in cases where these are permitted. They would not be allowed in drinking water reservoirs.

(f) The surface finish is somewhat rougher than *in situ* concrete, and this may not always be acceptable. The finish can be improved by careful trowelling. Care and experience in trowelling is particularly important, otherwise the plastic gunite may be displaced.

The UK Code for liquid retaining structures, BS 5337, Clause 28, states that pneumatically applied mortar may be used as cover to external prestressing wire, as rendering, and as structural concrete. The previous Code did not include the use of gunite for structural concrete. There seems no reason why gunite should not now be accepted and designed in the same way as concrete, using possibly Grade 30 or in special cases even Grade 40, in Table 7, page 10 of the Code. The practical problem is quality control and testing and this is discussed later in this section.

The wording of Clause 28 allows volume batching and the use of lime in the mix for pneumatically applied mortar for the cover coat to prestressing steel. The authors certainly do not agree with this, and consider it would have been better to follow the requirements of the British Standards for prestressed concrete pipes, BS 4625 and BS 5278. The use of lime will increase the permeability of the mortar and there is a danger of particles of unhydrated quicklime (CaO) being present, and this can cause a 'blow-out'.

As far as the authors are aware, none of the specialist contractors in the UK uses lime in the mix. Weigh-batching is recognized as being essential for uniformity in a concrete mix and this also applies to mortar mixes.

The use of lime was probably intended to provide a more workable mix and one less prone to shrinkage cracking. However, structural gunite has a low water/cement ratio, usually between 0·30 and 0·35. If this is considered too harsh, a normal plasticizer can be used; an alternative, but one which does not meet with general favour with gunite contractors, is to incorporate an artificial rubber latex, based on styrene butadiene or acrylic, in the mix. The use of such a latex will reduce drying shrinkage (by acting as a powerful plasticizer) and will also reduce permeability and increase resistance to chemical attack to a limited extent. However, the mix is 'sticky' and causes difficulty in gunning. The cement content is about 450 kg/m^3.

There are two guniting processes in general use, the dry-mix process and the wet-mix process. Of these, only the dry-mix process is in general use in the UK for normal gunite work.

In the dry-mix process the cement and sand should be weigh-batched and properly mixed, and then it is blown down the delivery hose by compressed air. The water required to hydrate the cement is delivered to the mix through a manifold (special mixing device) immediately before the nozzle of the gun. The amount of water, which is delivered to the manifold under pressure, is controlled by the operator (see Fig. 9.9).

In the wet-mix process, the cement, sand, and water, are weigh-batched and mixed in the normal way and then forced down the delivery hose by compressed air and projected on to the surface to be gunned. The advantage claimed for this is that all constituents of the mix, including the water, are weigh-batched and therefore any error by the nozzle man in judging the amount of water is eliminated. While this claim is undoubtedly valid and conforms with the basic requirement for high quality concrete, the fact is that consistent high quality gunite can and is produced by the dry-mix method, and satisfactory results are not always obtained by the wet-mix method.

Whichever method is used, the mortar when it is discharged from the nozzle must be well mixed and cohesive and capable of supporting itself on vertical surfaces and, when necessary, on overhead surfaces as well.

The properties and performance of gunite depend to a large degree on the skill of the operator, but grading of the sand, mix proportions, type of equipment used, and site conditions, are also important.

Generally the mortar in place is somewhat richer than the original mix proportions due to loss of sand by rebound. Figures for rebound vary, but

Fig. 9.9. *Guniting wall of static water tank.* (*Courtesy: Cement Gun Co. Ltd, London.*)

the following figures can be considered as reasonable for most jobs; they are taken from US experience:

Walls: 15–30%
Floors: 10–15%

The reinforcement should be designed and located so that maximum structural advantage is obtained, consistent with adequate cover. It should be detailed so that it causes as little obstruction as possible to the placing of the gunite. The reinforcement can consist of any of the types of steel used in reinforced concrete, but many specialist contractors prefer to use a heavy steel fabric to BS 4483—*Steel Fabric for the Reinforcement of Concrete*. The various fabrics set out in Table 1 of this Standard will provide adequate space between the main and cross wires to allow proper placing of the mortar/concrete. When reinforcing bars are used instead of the fabric, it is advisable that the distance between two parallel bars should not be less than 65 mm.

Execution of the Work

One of the most difficult but at the same time most important matters in gunite construction is to ensure that the gunite is of uniform density throughout, and that there are no hollow areas or sand pockets left behind reinforcement. Corners require special care, and when reinforcing bars are lapped it is advisable for the bars to be spaced at least two diameters apart to ensure that they are fully embedded in the mortar.

Experienced gunite contractors have various methods of eliminating sand pockets behind reinforcing bars, the principal one being 'piping'; this consists of blowing out the loose material which collects behind the bars as the work proceeds. This should be encouraged and checks carried out to ensure that sand pockets are not being formed.

From the remarks above it will be clear that the nozzle man is much more important in a guniting operation than any single member of a concreting gang.

It has been previously mentioned that the British Code, BS 5337, recommends a minimum cover of 40 mm of gunite over prestressing wires. While this may be justified for the protection of the small diameter, highly stressed wires in a circumferentially prestressed tank, the authors consider that 30 mm is adequate cover for normal reinforced gunite work. Cover in excess of 30 mm is undesirable as crack control becomes more difficult as the depth of cover is increased. If cover greater than 30 mm is provided, consideration should be given to the inclusion of a light galvanized steel fabric.

The order in which the various parts of the structure are gunited will depend on the shape and size of the structure. With open rectangular tanks it is usual for the walls and the adjoining bay in the floor to be 'gunned' in one operation. The walls are gunned in panels for their full height, and generally for their full thickness. This means that to a large extent the length of the panel is governed by the output of the equipment and the operator.

Unless it is otherwise specified, the walls are usually gunned against hessian which is stretched between timber profiles placed at about 1·0-m centres. Prior to the application of the gunite the hessian is given a coat of gunite or cement grout to stiffen it. However, the experience of the authors is that considerable distortion of the hessian and the timber frames does sometimes take place. This results in excessive cover which means that the reinforcement is not located where the designer intended it to be. In extreme cases this can result in a structurally unstable wall.

In view of the above it is recommended that plywood forms be used and that the supporting timbers be of adequate section to ensure rigidity.

Generally, gunite designers/contractors provide a wide cove angle at the junction of the wall and floor, of about 1·0-m radius. However, gunite structures can be designed and built with a normal right angle junction, and this has been done in a number of cases.

The floor is usually formed of large shingle, known as 'rejects', which is laid to a depth of 150–200 mm. The reinforcement is laid on this and the floor is gunned after the completion of the walls and perimeter strip. Other methods of forming the floor are sometimes adopted using concrete or a ready-mixed mortar. Care must be taken to ensure that the joint between the perimeter strip and the floor is properly made and is watertight. The authors feel that it is generally better if the structural floor slab is constructed of the same materials as the walls, i.e. reinforced gunite.

A point which requires attention is that there is a tendency for rubbish such as pieces of timber and cement bags, etc., to accumulate under the reinforcement of the floor slab prior to casting. This must all be cleaned out before any gunite or concrete is placed.

If a panel has to be increased in thickness after it has been gunned, the first layer must be allowed to take its initial set before the second layer is applied. All loose material should be carefully brushed off the surface. If the lower layer has already hardened, it is advisable for the surface to be lightly grit blasted and dampened with a water spray immediately prior to placing the second layer. The use of high velocity water jets, as described in Chapter 7, would combine the effects of cleaning, grit blasting and dampening.

It is unusual for any form of movement joints to be provided in gunite liquid retaining structures as is common practice in reinforced concrete structures. This is commented on in the next section. Construction (daywork) joints have to be provided because the structure cannot be gunned in one continuous operation.

There are differences of opinion among gunite contractors on the detailing of these monolithic joints. The ACI Code (ACI 506–66) recommends feather-edging for the full depth of the section (wall or floor). United Kingdom practice tends to favour a plain butt joint at right angles to the surface down to the reinforcement and then the gunite is tapered for the remaining depth of section.

An important matter which is sometimes neglected by even experienced contractors is the protection of the freshly placed gunite from strong winds and hot sun. Although the design of the gunite structure and the sequence of operations certainly reduces thermal contraction and shrinkage stresses, nevertheless the authors consider that gunite should be properly cured for at least 4 days after placing. Protection from wind and sun is of particular importance.

In the case of concrete structures, considerable reliance is placed on quality control of the mix, and on test cubes for controlling the strength of the concrete, and this subject is discussed at some length in Chapter 10. The quality control and strength testing of gunite is more difficult as neither cubes nor cylinders can be made in the usual way as laid down in BS 1881. With gunite structures testing should be done by either coring or by cutting samples from specially shot panels. The panels should be shot at the same time as the gunite is being placed in the structure. The taking of cores from the structure itself is obviously undesirable and should be avoided whenever possible.

Quality control of the gunite can only be exercised by weigh-batching and careful visual examination of the gunite as it is applied. Even so, as previously mentioned, the coefficient of variation is likely to be higher than with well controlled concrete.

Movement Joints in Reinforced Gunite

Normally, reinforced gunite water-retaining structures do not have any movement joints. They obviously have joints, but these are daywork joints and are monolithic. A natural question, therefore, is why should a reinforced concrete structure require movement joints in the form of contraction or stress-relief joints, while a gunite structure appears to be satisfactory without such joints? It is not easy to give a completely clear and unambiguous answer to this but the factors involved are as follows:

(a) Usually the walls of a gunite structure are thinner than a similar structure built in *in situ* reinforced concrete. This together with the absence of a normal formwork, at least on one side, results in a reduction in the build-up of heat and consequently lower thermal stresses in the walls than occurs in concrete walls which have formwork on both sides.

(b) Good quality gunite as placed has a very low water/cement ratio, probably about 0·33–0·35 with a resulting advantage of a cement matrix of low water/cement ratio. Concrete as used for most water retaining structures has a water/cement ratio in the range 0·45–0·50.

These factors add up to the fact that there is likely to be less thermal contraction and drying shrinkage in the walls of a gunite water retaining structure than in the walls of a similar structure in reinforced concrete.

The above remarks are not intended to imply that cracking is unknown in gunite water retaining structures. It does sometimes occur, but experience suggests that this is rather less frequent than in concrete structures. When

cracks occur, they are usually in the form of fine surface cracks a
to inadequate curing. The authors would like to make it clear tha
consider that even specialist contractors frequently do not take sufficie
precautions, particularly under adverse weather conditions, to prevent a too rapid drying out of the surface of the gunite, with the result that surface cracks appear. It is therefore recommended that specifications for gunite water retaining structures should contain provision for covering up (protecting) the gunite as the work proceeds and this should be enforced on site.

9.3. *Water Towers*

Introduction

Water towers are usually constructed of *in situ* reinforced or prestressed concrete and generally consist of an elevated tank supported on columns or on a massive central stem. The older, plain, square or rectangular concrete tanks on four columns have in recent years given place to very elegant and attractive structures.

In some instances, particularly on the Continent and overseas, the tank is cast at ground level and then jacked up the central stem.

Figures 9.10 and 9.11 show prestressed and reinforced concrete water towers constructed in the conventional way. Figures 9.12–9.15 show a water tower at Riyadh, Saudi Arabia, which holds 12 000 m^3 of water. The photographs show the various stages of construction. The tank was cast at ground level and then jacked up the stem. The water compartments were finished inside with two coats of a two-pack epoxide resin using a polyamide hardener.

Figure 9.16 shows a water tower near Liverpool. The shaft was slip-formed and then the tank was cast at ground level and raised to the top of the shaft by special jacks.

While the basic principles of construction which have been detailed in previous chapters are also applicable to water towers, there are two special features which will be discussed here. These are external appearance and the need for complete watertightness. Both are to some extent connected as even slight leakage is likely to have a serious effect on the appearance.

External Appearance

Water towers are normally conspicuous structures as they are located on relatively high ground and in addition they are elevated.

Fig. 9.10. *Prestressed concrete water tower, 18 m high (Courtesy: Preload Ltd, London.)*

Fig. 9.11. *View of reinforced concrete water tower under construction.*

This book does not cover the aesthetics of design of structures and some people will no doubt consider that this is the province of the architect rather than the engineer. It is suggested that good engineering should include a design which is aesthetically satisfying. The great engineers of the past have always achieved this. One of the many advantages of concrete as a construction material is the variety of external finishes which can be obtained.

Fig. 9.12. *Water Tower in Riyadh. Central shaft under construction, with reinforcement for tank being fixed. (Courtesy: Dyckerhoff & Widmann, München.)*

Fig. 9.13. *Water Tower in Riyadh. Tank being jacked up central shaft.* (*Courtesy: Dyckerhoff & Widmann, München.*)

When considering the type of finish, the problems of weather staining must be given close attention and careful detailing is needed if an otherwise pleasing structure is not to be spoilt by unsightly streaks and patches.

A great deal of work has been done in recent years on another type of concrete civil engineering structure, namely bridges. As a result, modern highway bridges are now being designed and constructed which are not only attractive when they are new but will remain so for a long period.

There has been a move away from smooth, plain finishes to a rougher type of surface, such as board-marked, exposed aggregate and sculptured.

Detailed advice on these finishes and practical examples of them can be obtained from the Cement and Concrete Association, at Wexham Springs,

Fig. 9.14. *Water Tower, Riyadh. View during second stage of jacking the tank. (Courtesy: Dyckerhoff & Widmann, München.)*

near Slough. Reference can also be made to the bibliography at the end of this chapter.

When considering the type of finish required, it is important to remember that the mix design chosen must be suitable for watertightness as well as external appearance. The type and grading of the aggregates, the proportion of fine to coarse aggregate, the water/cement ratio, the workability and method of placing the concrete, are essential parts of the concreting technique; this is why expert practical advice on how to achieve a particular finish is necessary early in the design process.

As an example, it is often very convenient to pump the concrete from either the point of discharge of the mixer or ready-mixed concrete vehicle, to the formwork. Pumped concrete requires special mix design and it may be found that the design for satisfactory pumping will not give the required surface appearance when the forms are struck. If marine-dredged

Fig. 9.15. *Water Tower, Riyadh. View of completed structure.* (*Courtesy: Dyckerhoff and Widmann, München.*)

aggregates are to be used, the effect of shell in the aggregate must be considered; also, salt can cause efflorescence. Pyrite is present in flint gravels from certain areas, and its presence will cause brown stains in isolated areas on the concrete. These look like rust stains and their appearance after a few years can cause consternation until the real cause is ascertained.

Watertightness

Nothing spoils the appearance of an otherwise attractive water tower more than seepage of water through cracks, joints, and defective concrete.

It is not a simple matter to design and construct a large tank at a height of

Fig. 9.16. *Reinforced concrete water tower with tank being jacked. (Courtesy: British Lift Slab Ltd, Birmingham.)*

Fig. 9.17. *Polyisobutylene sheeting to floor, columns and walls of water tower. (Courtesy: Gunac Ltd, Wallington, Surrey.)*

20–40 m above the ground, so that it is literally 'bottle' tight. A lower standard of watertightness can cause disappointment from an appearance point of view.

It is impossible to construct a structure of any size without joints and in a water tower these will be visible. Therefore they should be featured and form part of the external design, and must be completely watertight.

Fig. 9.18. *View showing detail at foot of reinforced concrete column in water tower covered with polyisobutylene sheeting bonded to base concrete.* (*Courtesy: Gunac Ltd, Wallington, Surrey.*)

The opinion has been expressed that circular concrete structures are particularly suitable for prestressing, and that prestressed concrete is in compression and this overcomes its basic weakness, namely low tensile strength. Therefore consideration can usefully be given to the use of prestressed concrete for the tank, even though this may not be used for the supporting structure.

If site conditions are such that prestressed concrete can be used for the tank, then the authors feel that such a tank should not need a lining. However, with *in situ* reinforced concrete, the provision of a very durable, watertight lining should be given careful consideration.

The lining can be either preformed material, such as PVC or polyisobutylene (PIB), or *in situ* coatings based on polyurethane and epoxide resins.

For towers containing potable water, the lining must be non-toxic and non-tainting, and must not support the growth of bacteria, fungi and algae.

With both types of material, it is essential that there should be maximum bond between the base concrete and the lining. This can best be achieved by light exposure of the coarse aggregate, followed by removal of all dust and grit. The lining must have the degree of flexibility needed to accommodate the structural and thermal movement which will take place in the tank during operation.

Generally, sheet material should not be less than 1·0 mm thick, and *in situ* coatings not less than 0·75 mm thick. The joints in the sheets should be solvent welded and not joined with an adhesive.

With coatings, there should be a minimum of two coats, and preferably three. These can be applied by airless spray or brush. If by brush, then each subsequent coat should be applied at right angles to the preceeding one in order to avoid pin holes.

Sheet material and coatings must be carried up above top water level; this is particularly important with sheet materials in order to avoid possible deterioration of the adhesive used to bond the sheets to the base concrete.

Figures 9.17 and 9.18 show the inside of a water tower lined with polyisobutylene, bonded to the walls, floor and columns.

9.4. *Unreinforced Concrete for the Walls of Water Retaining Structures*

It is usually assumed that a reinforced concrete wall of an 'average' type of water retaining structure costs less than a similar wall constructed in plain (unreinforced) concrete. However, investigations in the UK by one of the large Water Authorities in consultation with the Cement and Concrete Association have shown that this assumption is not necessarily correct in all cases. A detailed examination of all the factors involved proved that in many instances the overall cost of the plain concrete wall was less than that of the much slimmer reinforced wall. In addition, there are certain other advantages.

The most vulnerable parts of reinforced/prestressed concrete water retaining structures are the following:

(a) The joints; these are vulnerable to leakage.
(b) The joint sealants; those in general use are rubber-bitumens, polysulphides and silicone rubbers, and these deteriorate and have

to be renewed. The period depends on the original quality of the sealant and the characteristics of the liquid stored.

(c) The steel reinforcement which includes prestressing wire and strand. This is vulnerable to corrosion if the passivity provided by the cement matrix is for any reason reduced. This condition can occur due to inadequate cover to the steel, cracks allowing ingress of water, the presence of chlorides in the concrete mix or in the stored liquid.

(a) *Joints*. It is impossible to eliminate these, but they can be reduced in number and therefore in total length by increasing the weight of reinforcement used to control thermal contraction cracking (see Chapter 7).

(b) *The joint sealants*. The flexible joint sealants which exhibit the best durability characteristics are Neoprene and EPDM (ethylene–propylene–diene–monomer), but these are normally available only in the form of preformed gaskets. These gaskets can provide a watertight joint against quite considerable head, but they require an accurately formed sealing groove and this may involve reforming the sides of the groove with epoxide mortar. However, as discussed in Chapter 7 it may be possible to eliminate the use of a flexible sealant or at least reduce the degree of flexibility considerably, and then an epoxide mortar can be used to seal the joints which should prove almost as durable as the concrete itself.

From the discussion in (a) and (b) above, it can be seen that very little economy if any can be effected by change in the design or detail of the joints.

(c) *The steel reinforcement*. The cost of steel has increased very considerably in recent years, more so than almost any other constructional material. Very little can be done to decrease or eliminate the natural risks to its long-term durability, without incurring considerable additional expense in the form of watertight coatings to the concrete, etc. It is admitted that experience has shown that the risk of deterioration of reinforced concrete water retaining structures due to corrosion of the reinforcement is small. Nevertheless, every year there are a number of such structures which require expensive repairs due to this cause.

Therefore if steel reinforcement can be eliminated from a substantial part of

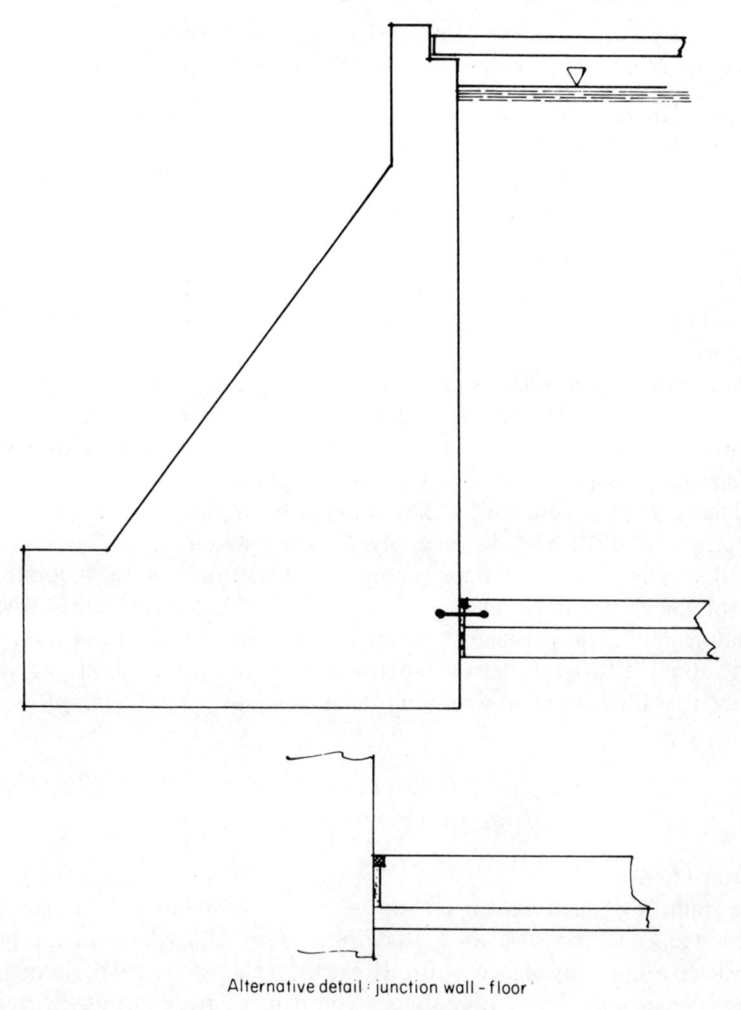

Fig. 9.19. *Plain concrete wall for reservoir.*

the structure (the walls), then in addition to the financial saving, there is likely to be an improvement from the long-term durability and maintenance aspects.

The plain concrete walls are designed on the gravity wall principle, the mass of the wall ensuring its stability against sliding and overturning. It has been found that the spacing of vertical joints can be at about 4·5-m centres which is similar to a conservative design for a reinforced concrete wall. The concrete mix must be carefully designed to achieve the following:

(a) The mature concrete must be durable and impermeable.
(b) The temperature rise in the maturing concrete must be as low as practical as this will reduce the chances of thermal contraction cracking during the first 20–72 h after casting.

Figure 9.19 shows such a wall which would have a panel length of 4·5–5·0 m. Each wall panel, complete with base, should be cast in one continuous operation; it would have a volume of about 100 m^3 and a mass of about 240 tonnes.

Dumb-bell section PVC water bars should be used in all vertical joints in the walls and the external type used for the joints in the floor slab. The joints between the base of the wall and the perimeter bay of the floor slab should be specially detailed so as to ensure watertightness.

The use of plain concrete for the walls greatly simplifies the design and the requirements of BS 5337 do not apply. The floor should be reinforced in the usual way in order to reduce the number and total length of joints to reasonable proportions. The roof can be a flat reinforced concrete slab with a sliding joint at the perimeter where it is supported on the walls. A saving of up to about 10 % in total cost of the structure can result from the use of plain concrete walls compared with reinforced concrete.

9.5. *Reinforced and Prestressed Concrete Tanks for the Storage of Petroleum Products*

Introduction

The traditional material for oil storage tanks throughout the world has been steel, and 'tank farms' as they have been called have in the past consisted exclusively of steel tanks. However, a change is now beginning to take place and international oil companies are starting to give consideration to the use of both reinforced and prestressed concrete tanks. There are many reasons for this and the more important ones are summarized below:

(a) The very large increase in the price of steel compared with concrete.
(b) The popularity of concrete and its proven value in all types of civil engineering and building structures.
(c) The need to give more consideration to the cost of maintaining the oil storage installations. Steel tanks require initial painting and

regular maintenance and repainting schedules; this applies to both the inside and outside of the tanks.

(d) Concrete tanks require no external coating; if the internal oil resistant membrane is correctly selected it should last almost indefinitely. The membrane is required to ensure complete oil-tightness and not to protect the concrete.

(e) Due to 'urban guerrilla' and other types of terrorist activity, many steel tanks are now protected by concrete bund walls (see Figs. 9.8 and 9.9). Concrete tanks are themselves very resistant to bombs, bullets and explosives.

(f) The environmentalist lobby now exerts considerable pressure and corresponding influence to prevent large tank farms disfiguring the countryside. One solution to this problem is to put the tanks wholly or partly underground. In such circumstances, concrete becomes very competitive in first cost with steel, and as stated in (d), maintenance is minimal.

(g) Reinforced and prestressed concrete tanks are much more resistant to fire than steel, and therefore should a fire start, the period which will elapse before collapse will be very much longer than with steel tanks.

Structural Design

The principles of structural design laid down in National Codes for water retaining structures can in general be applied to oil storage tanks.

Because the danger from any leakage is likely to be much more serious than with water retaining structures, the authors consider the use of the older elastic design principles are preferable to limit state design, even though the majority of the tanks will be lined. This is a matter on which the engineer should use his own judgement, taking into account all relevant factors.

Specification

The concrete should be high quality and the detailed mix proportions and other relevant requirements should be as recommended previously in this book, with special reference to Chapters 6 and 8.

It is recommended that on completion the tanks should be tested with water as described in Chapter 10. All leaks which show on the water test and all visible signs of seepage and sweating should be carefully repaired. General information on repair techniques is given in the book *Concrete Structures: Repair, Waterproofing and Protection.*

On completion of all repairs and making good, the whole of the inside of the tank should be provided with an oil resistant and durable membrane. This membrane should preferably be slightly flexible so that it will not fracture in the event of slight movement in the tank floor and walls. If the tank is designed so that there is a sliding joint between the walls and the floor (as is usual with prestressed circular tanks), then it is essential for the membrane to possess sufficient flexibility to accommodate this inward and outward movement. It is advisable for the membrane to be supplied and applied by one contractor, who should also be responsible for the preparation of the concrete to receive the membrane. In this way divided responsibility will be avoided.

Bibliography

American Concrete Institute. Design and construction of circular prestressed concrete structures. Title 67–40. *J. ACI*, **6**(9), September 1970, pp. 657–672.

American Concrete Institute. Guide to durable concrete. Report of ACI Committee 201. Title 74–53. *J. ACI*, December 1977, pp. 573–609.

American Petroleum Institute. *Evaporation Loss in the Petroleum Industry, Causes and Control.* API Bulletin No. 2613, Dallas, 1973.

Anon. *Notes on the Preparation and Filling of Ducts in Prestressed Concrete Members.* Code 47.012, Cement and Concrete Association, London, 1971, 5 pp.

Closner, J. J. and Porat, M. M. *Leakproof Concrete Tanks for Aviation Fuel.* Civil Engineering, USA, April 1960.

Comité Européen du Beton and Fédération Internationale de la Precontrainte. *International Recommendations for the Design and Construction of Concrete Structures.* Code 12.035, Cement and Concrete Association, London, 1970, 80 pp.

Concrete and Constructional Engineering. A 3 000 000 gal. reservoir with precast prestressed walls. *Concrete & Constructional Engineering*, July 1960, pp. 279–282.

Federation Internationale de la Precontrainte. *Recommendations for the Design of Prestressed Concrete Oil Storage Tanks.* FIP Task Group on Oil Storage Vessels, FIP/3/2, January 1978, 44 pp.

Flint, A. R. and Low, A. E. The construction of hyperbolic paraboloid type shells without temporary formwork. *Bull. Intern. Assoc. Shell Structures* (4), 1960.

Institute of Petroleum. *British Refining Safety Code*, 1965.

Leonhardt, F. *Prestressed Concrete Design and Construction*, Third Edition. Wilhelm Ernst, 1973.

Lin, T. Y. *Design of Prestressed Concrete Structures.* John Wiley & Sons Inc., 1955.

Low, A. E. The concept of an engineering design. Paper on the shell roof of the Baha'i Temple in Panama. *Concrete*, November 1972, pp. 24–26.

Matte, Waton & Rawlings. *Permeability and Properties of Oil Soaked Concrete.* Report No. R.53, University of Sheffield, Department of Civil and Structural Engineering Research, January 1970.

National Fire Protection Association (USA). *Flammable and Combustible Liquids Code.* Publication NFPA 30, 1973.

Perkins, P. H. *Concrete Structures: Repair, Waterproofing and Protection.* Applied Science Publishers, London, 1977, 302 pp.

Perkins, P. H. *Swimming Pools*, Second Edition. Applied Science Publishers, London, 1978, 398 pp.

RILEM–FIP–IABSE Committee. *Corrosion Problems with Prestressed Concrete.* Fifth Congress of FIP, Paris 1966, 6 pp.

Sawtell, D. L. *Concrete in Architecture.* Code 48.009, Cement and Concrete Association, London.

Schupack, M. Prestressed concrete tank performance. *Paper No.* 5, *Symposium on Concrete Construction in Aqueous Environments.* Publication SP 8, American Concrete Institute, September 1962, pp. 55–65.

Sommer, P. Liner systems for oil tanks. *IASS Meeting, San Diego, California*, June 1976.

Tiller, R. M. and Ward, F. *Concrete Finishes for Highway Structures.* Code 46.001, Cement and Concrete Association, London, 25 pp.

Trakern, J. W. Prestressed concrete tanks, design construction and maintenance. *Paper No.* 4, *Symposium on Concrete Construction in Aqueous Environments.* Publication SP 8, American Concrete Institute, September 1962, pp. 43–54.

Wilson, J. G. *White Concrete with Some Notes on Black Concrete.* Code 48.010, Cement and Concrete Association, London, 1969, 30 pp.

Wilson, J. G. *Specification Clauses Covering the Production of High Quality Surface Finishes.* Code 47.010, Cement and Concrete Association, London, 1970, 14 pp.

Wilson, J. G. *Abrasive Blasting.* Code 47.008, Cement and Concrete Association, London, 1971, 16 pp.

Winkler, N., Aschwanden and Speck, *Report on the Feasibility and Economical Merits of Prestressed Concrete Tanks for Oil Storage, Zurich*, 1976.

CHAPTER 10

QUALITY CONTROL AND TESTING

10.1. *Introduction*

The watertightness of liquid retaining structures is checked by means of a water test before final commissioning. However, if the design is based on the limit state principles set out in the Code (BS 5337), the width and spacing of the 'permitted cracks' cannot be checked under working load conditions, except the floors of elevated structures.

In addition to this final testing, specifications normally insist on a high degree of quality control of the concrete as well as close supervision of the execution of the work. A great deal has been written about quality control of concrete and the testing associated with it. For readers who wish to pursue this subject in depth, reference should be made to the bibliography at the end of this chapter.

The authors' object here is to emphasize certain practical considerations relating to quality control, summarize the more usual methods of testing hardened concrete in the structure and give recommendations for the final water test.

10.2. *Quality Control of Concrete*

Cube Testing

One of the basic principles of quality control procedures is that these are based on statistical methods and the mathematical laws of statistics. When a large number of cubes are tested it is statistically inevitable that a certain number will fall below the specified minimum strength; the actual number depends on many factors and this is dealt with in the UK in Code of Practice CP 110—*The Structural Use of Concrete* and elsewhere in the relevant

National Codes. Generally a failure rate of 5% is considered reasonable for good quality structural concrete. This means that the average strength required to meet such a specification must exceed the specified (characteristic) value by a factor taken as 1·64 times the standard deviation.

The object of quality control is to ensure that the concrete produced complies consistently with the requirements of the specification within the statistical limits prescribed. If the batching, mixing, transporting, sampling and testing are all under good control and carried out correctly, then concrete of consistent quality with a small standard deviation (current margin) will result.

The general practice of judging the quality of concrete on the result of testing 100-mm or 150-mm cubes is criticized on a number of grounds, all of which are valid as far as they go. Unfortunately, no one has been able to come forward with a practical alternative suitable for use on construction sites. The fundamental objection to test cubes is that this method merely gives information on the strength of the concrete in the cubes which are made from the concrete before it is placed in the formwork; therefore the test cube results may bear little relation to the strength of the concrete in the structure. Test cubes have to be prepared and tested in accordance with the detailed requirements of BS 1881—*Methods of Testing Concrete*. The requirements of the Standard are intended to provide a clear procedure, so that apart from human and test equipment error, all cubes are processed in exactly the same way. The history of the concrete in the structure, from the time of sampling to the time it is, say, 28 days old, is obviously quite different to that of the concrete in the test cubes. The usual method of testing concrete in the structure is to take 100-mm diameter- or 150-mm-diameter cores and test them in accordance with the relevant clauses in BS 1881. Some remarks on core testing are given later in this section.

In many instances when disputes arise on site because the test cube results are below those specified, the case is lost because investigation shows that the sampling and/or testing procedure was in some respects faulty due to the fact that the detailed directions laid down in BS 1881 were not strictly adhered to. It is therefore essential to ensure that the whole of the sampling and testing is done in accordance with the British Standard; this also means that all equipment used is in proper working order and regularly checked for accuracy.

About 50% of all concrete used on construction sites in the UK is now supplied by the ready-mixed concrete industry. In the USA the figure is about 75%. The authors consider that if a contractor wishes to use ready-mixed concrete he should be allowed to do so subject to certain conditions which should be discussed and agreed between the resident engineer, the main contractor and the concrete supplier, and finally confirmed in writing. These conditions should be based on the following:

(a) The resident engineer or one of his senior staff should visit the concrete depots from which it is intended to supply the concrete, and satisfy themselves that the depots are in fact able to produce consistently the quality and quantity of concrete required for the job. The depots should be registered under the British Ready Mixed Concrete Association's Authorization Scheme, preferably for quality control procedures.
(b) Agreement should be reached between the parties on the interpretation of the clauses in the specification for the concrete.
(c) The concrete should be supplied in accordance with BS 1926—*Ready-Mixed Concrete*. The resident engineer, his staff and the main contractor should be acquainted with the provisions of this Standard.
(d) The ordering procedure for the concrete should be agreed and adhered to; BRMCA have clear and detailed recommendations for this.
(e) A procedure should be worked out for joint sampling and testing of the concrete; all parties concerned should then accept the test results as relevant to the quality of the concrete as it is delivered to site.

It is not claimed that the proposals outlined above are easy to put into effect, nor that they will necessarily eliminate all disputes over the quality of the concrete; but if there is common sense on both sides it will go a long way towards that objective.

A serious problem can arise with cube testing due to the fact that most specifications refer to strengths at 7 and 28 days. If the 7-day results are low but the 28-day strengths comply with the specification, then the concrete is normally accepted as satisfactory. This means that nothing much can be done until 28 days have elapsed after casting the concrete, and in that period an appreciable amount of additional concreting may have been carried out. To overcome this, efforts have been made for many years to develop an acceptable technique for accelerated curing of concrete cubes so that the short period test results can be realistically and consistently related to standard 28-day strengths. The various methods so far put forward have not found much favour and have not been generally accepted.

Rapid Analysis of Fresh Concrete

The problem is particularly acute when using ready-mixed concrete as the batching and mixing is not under the direct supervision of the site supervisory staff. Efforts were therefore directed towards finding a method for checking some essential quality of the concrete when it is delivered to site before it is actually placed in the formwork. One essential factor of concrete is its cement content; if this could be measured in plastic concrete with a reasonable degree of accuracy and quickly (say in about 5–10 min) this would be very advantageous. A number of organizations carried out research with the object of developing such a piece of equipment. The first commercially viable equipment was developed by the Cement and Concrete Association and has been tested and used on many construction sites with satisfactory results. The equipment measures, in a period of a few minutes the weight of material contained in a weighed sample of plastic concrete which has approximately the same particle size as Portland cement. This means that if, for example, PFA (which has a similar specific surface to that of cement) were added to the mix and the fact was not disclosed, then the equipment would not distinguish between the cement and the PFA, although an experienced operator may notice certain things about the mix which would make him suspicious.

Site experience has shown that the accuracy for assessing the cement content is likely to be about $\pm 7\%$. This means that for a specific minimum cement content of, say, 350 kg/m^3, the equipment may record 326 kg or 375 kg. This would be a useful and practical piece of information, and should in most cases enable site supervisory staff to accept or reject concrete on the basis of the cement content before it is placed in the forms, provided the specification and the order to the supplier of the concrete clearly laid down the minimum cement content and prohibited the use of admixtures without prior written approval by the engineer.

In this book the recommended specifications for all structural concrete include a minimum cement content, maximum water/cement ratio, and generally a characteristic strength compatible with the first two requirements and taking some account of the type of aggregate used; in other words, the strength requirement is related to the quality of the concrete as specified and not to the strength used in the structural calculations. The reason for this is that in liquid retaining structures the maximum permitted design stresses are comparatively low while the requirements for durability and impermeability are high. Reference should be made to the section in Chapter 6 on durability and impermeability (p. 131).

10.3. *Testing Concrete in the Structure*

There are three main methods of testing concrete and checking reinforcement in the structure:

(a) Non-destructive testing:
- (i) the Schmidt rebound hammer;
- (ii) gamma radiography;
- (iii) cover meter for checking position of steel reinforcement.
- (iv) ultra-sonic pulse velocity;

(b) Core testing.

(c) Loading tests.

Non-destructive Testing

Each of the four techniques mentioned above is a subject in its own right, and quite a lot has been written about each. Readers who wish to obtain detailed information should refer to the bibliography at the end of this chapter, as only a few general notes will be given here.

THE SCHMIDT REBOUND HAMMER

This is a useful and practical instrument for site work, but the results must be interpreted with discretion, taking into account its limitations.

The instrument measures the surface hardness of the concrete, but experience shows that this can be related empirically to the compressive strength, provided certain precautions are taken. These may be summarized as follows:

(a) The instrument should be calibrated on the type of concrete being tested; 150-mm cubes are better than 100-mm cubes. It is necessary to recalibrate for cements other than Portland; for example, concrete made with high alumina cement may give 100% higher strength than similar concrete made with ordinary Portland cement.

(b) A minimum of 15 readings should be taken on each structural unit.

(c) The highest and the lowest reading should be eliminated and the average taken of the remainder. The average rebound number is then referred to the calibration chart and the strength read off. Alternatively, each of the readings, except those rejected, can be converted to compressive strength, and then these strengths are averaged.

It is useful to compare the readings on concrete which have been accepted as satisfactory with similar concrete in similar structural unit(s) which has given cause for concern. The usual reason for this 'concern' is that the cube results were below the specified minimum. If the readings on the suspect unit(s) are significantly lower, then this would support the view that the concrete was in fact of lower compressive strength. However, the results are only indicative, and if it is decided to continue with the investigation, the next step would be to either take cores or carry out a UPV survey. It is of course assumed that the sampling and testing procedure for the cubes has been checked and found to be in order.

There is a British Standard, BS 4408—*Recommendations for Non-destructive Methods of Test for Concrete*, of which Part 4 gives recommendations for surface hardness methods of non-destructive tests for concrete.

Gamma Radiography

This method of inspection of hardened concrete is suitable for thicknesses up to about 450 mm. It should be noted that gamma radiography comes within the scope of the Factories Acts and the Ionising Radiations (Sealed Sources) Regulations 1961. It is used for checking the efficiency of the grouting in cable ducts in post-tensioned prestressed concrete, to detect voids (honeycombing) in concrete, and to determine accurately the position and size of reinforcement.

Recommendations for this method of non-destructive testing are given in BS 4408: Part 3. The work is expensive and requires specialist operators as well as experience in the interpretation of the photographs; also, it must be possible to have access to both sides of the structural units under investigation.

Cover Meter Surveys by Electromagnetic Cover Measuring Devices

This is a comparatively simple piece of equipment, and shows the cover to reinforcement on a graduated scale. On a construction site, the accuracy of a good commercial instrument is about ± 5 mm.

It is normal to calibrate it for mild steel and for use with Portland cement concrete, but it can be used satisfactorily with other types of cement and high tensile steel, provided it is specially calibrated for the purpose. Some types of stainless steel are non-magnetic and so this type of instrument will not detect the metal; in such cases the alternative is gamma radiography.

Change in bar diameter from, say, 10 mm to 32 mm will not appreciably affect the cover shown, but for small diameter bars of, say, 5 mm, there will be a considerable difference between the indicated cover and the true cover; the cover shown on the scale is likely to be greater than the actual cover.

The relevant British Standard is BS 4408: Part 1.

In checking for reinforcement in old concrete structures, say built before about 1910, attention should be paid to the type of aggregate used. The reason for this is that aggregate containing ferrous material was sometimes used and this can result in the cover meter showing the presence of 'reinforcement', when in fact none is there, and it is the ferrous material in the aggregate which causes the movement of the pointer.

Ultra-Sonic Pulse Velocity Testing

The authors wish to record that the information which follows is largely based on discussions with Mr H. N. Tomsett of the Cement and Concrete Association, London, and a reference to a paper by Mr Tomsett is given in the bibliography at the end of this chapter.

The basic principle of ultra-sonic testing of concrete in the structure is that concrete is an elastic material and will transmit longitudinal, compression, and shear waves. The velocity with which these waves travel through the concrete is determined by those properties of the concrete which control the elastic modulus. These properties are in turn related to the strength of the concrete. The apparatus used generates a pulse in the concrete by the application of a mechanical impulse, it collects the impulse at some point at a measured distance from the point of generation, and contains a timing mechanism which accurately measures the time taken for the leading edge of the pulse to pass from the transmitter to the receiver.

The apparatus mainly used in the UK is that developed by Elvery at University College London and produced by CNS Instruments Limited. It is given the trade name of Pundit (Portable ultra sonic non-destructive digital indicating tester). The apparatus is small and readily portable as it weighs only about 3 kg. There are three basic ways in which the pulse can be transmitted and recorded. Direct transmission is most satisfactory; in this case the time measured is that for the longitudinal compression wave to pass between the transmitter and the receiver. The transmitter and receiver are in this case on opposite sides of the structural unit concerned. The next method, which is not so satisfactory as the previous one, is the semi-direct which can be used for such units as thick floor slabs where access can be obtained to the top surface of the unit and to the sides but not to the underside. The third method, which is the least satisfactory, is indirect, where transmission and receiving has to be carried out from one side only, as in ground floor slabs and basement walls after backfilling has been carried out.

It should be noted that the pulse transit time and strength do not have a direct proportional relationship in a given concrete. The transit time is measured very accurately by the instrument and it is important that the path length be also carefully measured or calculated.

There is a British Standard BS 4408, of which Part 5 is *The Measurement of the Velocity of Ultra Sonic Pulses in Concrete.* In this book it is not possible to go into the details of ultra-sonic pulse velocity testing but those who consider using it (and there is no doubt it is an extremely useful method of tesing concrete in the structure) should bear in mind the factors which affect the pulse velocity through concrete; these have been listed by Tomsett as

(a) aggregate type,
(b) aggregate grading,
(c) cement type,
(d) cement content,
(e) water/cement ratio,
(f) degree of compaction of the concrete,
(g) curing temperature.

The presence of reinforcement will affect the velocity obtained, and it is therefore advisable to make every effort to ensure that the pulses in the concrete structure are transmitted between the reinforcing bars. If this cannot be done, the interpretation of results is more complicated.

When used by an experienced engineer, the 'pundit' can in most cases determine satisfactorily the depth of cracks and the location or extent of under-compacted, poor quality concrete. The authors feel that the ultra-sonic pulse velocity method of non-destructive testing is particularly applicable to water retaining structures because the taking of cores in such structures below the water line is something which should be avoided whenever possible.

Core Testing

Core testing of the concrete in a structure is generally the last but one resort when cube results are significantly below the specification and the concrete is seriously suspected for being below the required strength. It is expensive and time consuming, and the results may not be as clear-cut as one would wish.

The taking of cores in a liquid retaining structure should not be lightly undertaken, as it means boring holes 100-mm or 150-mm diameter through the structure and it is not easy to ensure that the subsequent repair is watertight.

The cores should be taken and tested in accordance with BS 1881. The Standard gives detailed recommendations for the preparation, testing and examination of cores, but the interpretation of results requires considerable experience.

If the coring, testing and examination is carefully carried out it will give a considerable amount of information about the quality of the concrete in the structure. On the basis of the core results, a decision can be taken as to whether the concrete is satisfactory or whether the section in question may have to be demolished, and this is a very serious matter. When the results are still not entirely clear, a load test may be ordered, but not for walls. In some cases, attempts are made to establish a correlation between the core strengths and the cube strengths. While it may be possible to establish a reasonable relationship under strictly controlled conditions, the authors do not consider that this can be done satisfactorily on normal construction sites. The cubes are usually taken at the point of delivery of the concrete from the ready-mixed concrete lorry, or near the site batching plant. The cubes are then made, cured and tested in accordance with BS 1881.

The history of the two lots of concrete (cubes and that in the formwork) is quite different. The cubes are dealt with in a detailed and meticulous way, while the bulk of the concrete is transported by skip or pump, dropped into formwork containing reinforcement, compacted by poker vibrators by men perhaps working in strong winds and driving rain.

Apart from what has been said above, the following also introduce variables between cube and core strength:

(a) The presence of reinforcement in the core.
(b) The difference in age; cores are usually taken later than 28 days after casting the concrete; see also remarks below.
(c) Depending on the position in the structure, the core may have dried out a little more on one side than on the other; a curing membrane would influence the drying out process.
(d) There may be a noticeable difference in compaction between one end of a core and the other, particularly in floor slabs; in cores taken at the bottom of a lift and near the top, as in walls and columns, the strength of the former may be 20% higher than the latter.

In view of the above, the authors consider it unrealistic to try to assess the quality of the concrete as it was delivered to site (or as it left the batching plant) on the basis of core test results. A core will indicate the type and quality of the concrete in the structure.

A further complication is the rate of gain of strength with age between the concrete in the structure and the test cubes. Codes of Practice, Standards, and Specifications normally assume that the rate of gain of strength for both concretes (cubes and structure) is the same. As far as the authors are aware, this vital assumption has only recently been queried. Work and investigations by W. E. Murphy of the Cement and Concrete Association, London, has raised serious doubts about the validity of this age–strength relationship in structures. Conclusions to date suggest that in fact there may be little increase in strength in the concrete in what may be termed a 'normal' building structure after about 28 days from casting. There is no doubt that test cubes, dealt with as prescribed in BS 1881, do gain in strength after the first 28 days, due to the difference in environmental conditions.

Load Testing

Load tests are only likely to arise for suspended floor and roof slabs and beams including supporting columns when all other methods of non-destructive testing which are practical to the circumstances of the case have failed to resolve satisfactorily the question as to whether or not the part of the structure in question is structurally sound, or for testing old or damaged structures.

This type of testing is covered by Section 9 of CP 110. For readers who prefer to keep to the original Codes (CP 114, 115 and 116), the relevant clauses are CP 114, Clause 605; CP 115, Clause 602; CP 116, Clauses 504 and 505.

The useful papers on the load testing of reinforced concrete structures, one by J. B. Menzies and one by D. J. Jones and C. W. Oliver, were published in the *Structural Engineer* in December 1978. In a paper by G. F. Blackedge in November 1964, the author drew attention to the considerable differences in the requirements of the British Code and the American Concrete Institute Standard 318–63 and the Australian Code for Concrete in Buildings As-CA2-1963.

Doubts as to the structural soundness of a structure or part of a structure can arise from a number of reasons, of which the following are the most usual:

(a) Failure of test cubes, followed by failure to resolve this by core testing and/or other non-destructive tests.
(b) An acknowledged error in design.

(c) An acknowledged or suspected error in construction.
(d) Cracking and/or excessive deflexion.
(e) An increase in operating load conditions not anticipated during design and construction.
(f) Damage to the structure caused by fire, impact, etc.

Basically the tests are of two types; one for serviceability such as deflexion under design loading, and the other for safety, in which the design load is increased by some factor.

Load testing is a tedious business and should only be resorted to when all else has failed and the responsible engineer is faced with a decision to either order the demolition of the structure (or part) or to accept a structure on which he has doubts about its strength.

10.4. *The Water Test*

It is usual for all liquid retaining structures to be tested with water before they are accepted as satisfactory. The Code of Practice BS 5337 indicates briefly the type of test which is considered suitable. The relevant clause is 32, and this suggests that a drop in water level in a closed structure (that is one with a roof) of 10 mm, including evaporation, over a period of 7 days can be considered generally satisfactory. The clause goes on to say that for large, open reservoirs special allowance should be made for evaporation. No indication is given on how evaporation should be allowed for.

The authors' recommendations for carrying out a water test for the usual run of liquid retaining structures are as follows:

(a) A careful inspection should be carried out of the structure to ensure that all joints have been properly sealed (where sealing is allowed for in the detailing of the joint) and also that all inlet and outlet pipes below top water level are properly closed off by valves or blank flanges.
(b) The structure should be filled slowly. A rate of filling of 2·00-m depth in 24 h is considered satisfactory in the Code, Clause 4.4, but the authors recommend a somewhat slower rate of 1·5 m in 24 h.
(c) When the structure has been filled in this way to top water level, this level should be recorded and the fall in level should then be recorded each day over a period of 7 days. This is termed the initial soakage period, and if the structure is basically watertight it will be found that the fall in level diminishes each day. Bearing in mind that the test will require a maximum fall in level of 10 mm over 7 days, the fall in level over the last 24 h of the initial soakage period should be

relatively small. If this is not the case then it is advisable to continue with the preliminary soakage for a further 7 days. The period given here as 7 days is only approximate, and will depend very largely on the length of time that the reservoir has been empty, and the climatic conditions during that period.

(d) When the engineer is satisfied that the soakage period has been completed, the test should be started by refilling the tank to top water level and accurately recording the level. This can be done in a variety of ways including a mark on the wall, and better still a hook gauge. The structure is then allowed to stand for a period of 7 days but it is advisable to check and record the level each day.

(e) It is also advisable to make a practical test on how much evaporation is taking place. The Metropolitan Water Board (now the Thames Water Authority) have found that even with closed reservoirs, evaporation can be appreciable; the actual figure depends on the ambient air temperature and the strength and direction of the wind in relation to openings in the roof when the reservoir is closed. A practical way of checking evaporation is to fix in the water in the reservoir, a steel tank filled with water to within a few inches of the top. Any drop in level in the steel tank would then be considered as evaporation loss.

The authors strongly recommend that this procedure be adopted in every case. The reason for this is that if the test is successful then no questions and no queries arise, but if it is not successful then the contractor is certain to raise the question of evaporation and usually claims that the excess drop in level is due to evaporation. In such circumstances the only way to overcome it satisfactorily is to retest and at the same time make the test for evaporation. This of course wastes a considerable amount of time.

If backfilling has not been placed around the outside of the structure then any leaks in the walls are readily apparent. When there are leaks they usually occur at the joints and sometimes through fine cracks within the wall panels. When the cracks, or the opening of a joint, are very small, such defects are often self-sealing. It is recommended that all such defects should be carefully marked on the outside of the wall, and then when the test is over measures should be taken to seal the defects on the inside.

However, if the test is not successful, and no measurable leaks can be seen around the walls, then it means that the leakage is occurring through the floor, provided of course that there is no water passing through pipelines which enter the reservoir below top water level. Leaks in floors are very difficult to find. The authors' experience is that these almost invariably occur at joints, and it is very seldom that measurable leakage occurs through the concrete itself unless there are serious cracks, which would be visible.

Regarding 'measurable leakage', the following figures have been taken from the excellent CIRIA Report No. 81, *Tunnel Waterproofing*:

3–4 drips per minute is about 1 litre per day.
A drip is when drops of water fall at a rate of at least one per minute.
Seepage consists of visible movement of a film of water across a surface.

A detailed discussion with recommendations on repairing leaks in water retaining structures is given in the book *Concrete Structures: Repair, Waterproofing and Protection.*

Bibliography

British Standards Institution. *Assessment of Concrete Strength in Structures*. Draft for public comment, Document 77/13782DC, December 1977, 20 pp.

Canadian Portland Cement Association. *Design and Control of Concrete Mixtures*, Metric Edition. Printed in USA by the Portland Cement Association, Skokie, USA, 1978, 131 pp.

CIRIA. *Tunnel Waterproofing*. Report No. 81, April 1979, 55 pp.

Forrester, J. A. Gamma radiography of concrete. Paper presented to the *Symposium on Non-destructive Testing, London*, June 1969; Cement and Concrete Association, London, Paper for Publication No. 53. February 1969, 19 pp.

Jones, D. S. and Oliver, C. W. The practical aspects of load testing. *Structural Engineer*, **56A**(12), December 1978, pp. 353–356.

Kolek, J. An appreciation of the Schmidt rebound hammer. *Magazine of Concrete Research*, **10**(28), 1958, pp. 27–36.

Levitt, M. The initial surface absorption test—a non-destructive test for the durability of concrete. *Brit. J. Non-destructive Testing*, **13**(4), 1970.

Menzies, J. B. Load testing of concrete building structures. *Structural Engineer*, **56A**(12), December 1978, pp. 347–353.

Murphy, W. E. The assessment of concrete strength in structures. Paper given at *RILEM Symposium Quality Control of Concrete Structures, Stockholm*. PP/223, Cement and Concrete Association, London, February 1979, 8 pp.

Perkins, P. H. *Concrete Structures: Repair, Waterproofing and Protection*. Applied Science Publishers, London, 1977, 302 pp.

Tomsett, H. N. Ultra sonic testing for large pours. Paper at the *Concrete Society West Midland Symposium on Large Pours, Birmingham*, September 1973, 4 pp.

CHAPTER 11

COMMENTS ON BRITISH STANDARD BS 5337—*CODE OF PRACTICE FOR THE STRUCTURAL USE OF CONCRETE FOR RETAINING AQUEOUS LIQUIDS*

11.1. *Introduction*

The comments in this chapter are intended to be constructive, and to bring out important matters in the Code which the authors feel should be given careful consideration by engineers.

The authors have served on a number of Code and Standard committees and can therefore claim to have reasonable knowledge of how such committees work. Decisions are taken by 'concensus', which means that when differences of opinion arise there may be prolonged discussion until agreement is reached on the wording. Should a real impasse occur, the Chairman may end the discussion and then the matter is referred to a higher committee for final decision. A vote is never taken.

This may sound rather unsatisfactory, but in fact, on the whole, it works well, and Codes and Standards are sound documents and represent the general opinion of a group of experienced professional men. It does, however, sometimes result in a certain amount of imprecision and lack of clarity in the final document.

11.2. *Detailed Comments on Some Specific Clauses*

CLAUSE 4.9—DEGREES OF EXPOSURE

The Code recommendations are shown in diagram form in Fig. 8.1 (p. 203).

The upper parts of the inner surface of the walls of covered reservoirs and tanks are in reality continually moist, and this also applies to the underside of the roof slab. The Code suggests that they are generally subjected to alternate wetting and drying.

The concrete on the water face in the section between top and bottom water level will in time become saturated in covered structures. However, storm tanks and balancing tanks are different and are likely to be subjected to alternate wetting and drying. It is not clear why these conditions in water and sewerage structures should require a higher standard of concrete than in the parts which are more or less continually immersed. Such a requirement is justified on technical grounds in the splash zone of marine structures and structures holding highly saline liquids which are emptied at frequent intervals. If this Clause is read with Table 8, page 11, of the Code, the inference is that different grades of concrete can be used in two halves of a wall or floor which is more than 225 mm thick, which the authors feel is rather impractical.

Under Class A exposure, reference is made to 'Exposed to . . . corrosive atmosphere . . .' This presumably means corrosive to concrete (and not to unprotected steel reinforcement). It is felt that in such cases, and they do occur occasionally, the concrete is likely to need a correctly formulated protective coating. One example is covered tanks and sumps containing septic sewage, when attack by sulphuric acid above the water line (formed from hydrogen sulphide by two stage bacterial action) may occur. This requires special treatment and cannot be satisfactorily dealt with by simply increasing the cement content. Some comments on such cases have been given in Chapter 6.

Section Two—Design: Objectives and General Recommendations and Section Four—Design, Detailing and Workmanship of Joints

The principle of allowing cracks having a calculated maximum width in the tension zone of structural concrete members is now accepted as normal design practice for buildings, and the revised Code has extended this to water retaining structures. The same principle of 'controlled and calculated crack widths' is accepted for cracks caused by thermal contraction.

Cracks in the tension zone of structural members are assumed to have their maximum width at the surface of the concrete and then reduce down to the reinforcement where they are much narrower, and they seldom penetrate beyond the main bars.

Thermal contraction cracks, on the other hand, are parallel walled cracks; they retain, more or less, the same width right through the full thickness of the member.

Recent research in Germany and elsewhere on the effect of crack width (that is, width at the surface of cracks in the tension zone), suggests that the actual width has little effect on the corrosion of the reinforcement at the base of the crack. However, the authors have seen no published work which claims that a concrete member containing tension cracks which extend down to the reinforcement is as durable as a member without such cracks.

The whole question of the long-term durability of concrete, with particular reference to the effect of cracks, is under continuous discussion and there are undoubtedly differences of opinion on the subject. The evidence is somewhat conflicting and the possible effect on reinforcement of the carbonation of concrete within the cracks has not been satisfactorily resolved.

Apart from the practical difficulty in measuring the width of cracks in order to decide whether the crack width is within the limits imposed by the Code, three questions arise:

(a) If it is decided that, for example, cracks caused by thermal contraction in a wall exceed the permitted width, which party to the contract is responsible for this.
(b) Whether the cracks should in any case be repaired, that is, sealed.
(c) If the cracks have to be repaired, then a decision has to be taken on which party to the contract should pay for the work.

The authors' opinion is that all cracks which are visible to the naked eye should be sealed.

Clause 13.1.2 permits the use of 'partial' contraction joints and Fig. 1(b) on page 16 of the Code shows such a joint. The whole question of joints has been discussed in detail in Chapter 7 to which the reader is referred. It will be seen that the authors do not advocate the use of partial contraction joints. If shear has to be allowed for, then dowel bars de-bonded on one side of the joint should be used, although shear is seldom a significant factor in the walls of water retaining structures.

Section Six—Specification and Workmanship: Concrete

Clause 21.1. Cements

High alumina cement is not included in the cements permitted under the Code and therefore cannot be used for structural concrete. Its use for rendering and non-structural gunite is open to question. Correctly proportioned HAC mortar is rather more resistant to some dilute acids and solutions of sulphates than Portland cement mortar.

Clause 21.2. Aggregates

In the Code, the permitted absorption of aggregates is now 3 % which is an increase of 50 % over the previous Code. Detailed comments on the absorption of aggregates and its relationship to permeability and hence durability of concrete have been given in Chapters 5 and 6. In view of the possible financial implications of the 3 % restriction, it is felt that some technical justification for it would have been useful. The recommendation that aggregates should exhibit 'low drying shrinkage' seems rather imprecise for a Code of Practice. The Building Research Station in East Kilbride have done a great deal of research on the shrinkage of Scottish aggregates and reference can be made to BRE publications on the subject.

Clause 22.3. Minimum Cement Content

Minimum cement contents for two classes of exposure, Class A and Class B, are given in Table 8, page 11, of the Code. Comments have been given on the Code recommendations for conditions of exposure. The authors doubt whether the differentiation in cement contents for these conditions of exposure is really justified for the majority of reinforced concrete liquid retaining structures. In general, this book recommends somewhat higher cement contents than those in the Code.

It is felt that in the majority of structures covered by the Code, a more practical differentiation of cement contents would be to base this on the thickness of the concrete section rather than the degree of exposure. For example, a thickness of 600 mm could be used as the dividing line; sections less than 600 mm thick could have a higher cement content than thicker sections. The selection of such a thickness limit is of course rather arbitrary, but it has the advantage of simplicity and precision, and the principle is technically sound.

It is noted with some regret that the Code contains no recommendations for maximum water/cement ratio for structural concrete. Concrete technology research contains indisputable evidence that water/cement ratio is a vital factor in determining the permeability and hence durability of reinforced concrete. A maximum free water/cement ratio of 0·5 has been generally recommended throughout this book. It should be remembered that the 0·5 is a maximum, and therefore efforts should be made to obtain the necessary workability with a lower figure.

Clause 28. Pneumatically Applied Mortar

From the wording of the clause, it appears that this material can be used for structural purposes as well as a protective cover coat to prestressing tendons and wires in prestressed concrete structures.

It is noted with some regret that volume batching is permitted by the Code as it is well known that weigh-batching is necessary to ensure uniform quality of the batched material.

Cement mortar used to protect prestressing tendons and wires against corrosion must be of the highest quality and this can only be achieved by strict control of the mix and good supervision.

In view of the need for maximum impermeability of the cement mortar cover coat, it is surprising that the Code permits the use of lime in the mix. Reference to the Standards which control the manufacture of prestressed concrete pipes (BS 4625 and BS 5178) will show that the cover coat to the prestressing wires must consist of 1 part of Portland cement to 3 parts of sand by weight. The authors would not recommend the use of lime in the cover coat to prestressing tendons and wires.

Section Nine—Inspection and Testing

CLAUSE 32. TESTING FOR WATERTIGHTNESS

The wording of this clause suggests that in liquid retaining structures, except large open reservoirs, the permitted drop in water level of 10 mm during the 7-day test period should cover all losses including evaporation. Significant evaporation can take place in covered structures under certain climatic conditions. The authors therefore feel it would be better to exclude evaporation from the permitted drop in water level during the test period, and to measure this separately. This is discussed in detail in Chapter 10.

Appendix D—Jointing Materials

CLAUSE D.5. JOINT SEALING COMPOUNDS

The last sentence states '... most sealants should be applied in conditions of complete dryness and cleanliness'. It is difficult to visualize how these rigid requirements can be met in the majority of concrete liquid retaining structures built in this country and overseas.

11.3. *Notes on the Handbook to the Code of Practice for Concrete Liquid Retaining Structures*

The object of a handbook on a Code of Practice is normally to explain and comment on the code requirements, particularly those clauses which are more complex than others. It is therefore surprising that the Handbook to

BS 5337 should go beyond this and suggest alternatives and/or modifications to the Code recommendations.

Examples include additional options for design to control thermal contraction cracking and modifications to the recommended conditions of exposure. Engineers using the Code will naturally wonder whether the use of these additional options and modifications in the Handbook can be considered as part of the code recommendations.

The Handbook does not mention that limit state design was introduced into the Code without previous experience in its use in this country for liquid retaining structures. There does not appear to be any practical way of checking the calculated crack widths on the water face of this type of structure under service load conditions.

APPENDIX 1

STANDARDS AND CODES OF PRACTICE

*A*1.1. *UK Standards and Codes of Practice Relating to Materials Used in Construction*

CEMENTS	
Portland Cement, Ordinary and Rapid-hardening	BS 12
Portland Cement, White and Coloured	BS 12
Portland Cement, Sulphate-Resisting	BS 4027
Portland Cement, Ultra-High Early Strength	None
Portland—Blastfurnace Cement	BS 146
Portland—Blastfurnace Cement, Low Heat	BS 4246
Portland Cement, Low Heat	BS 1370
High Alumina Cement	BS 915
Supersulphated Cement	BS 4248
Pozzolanic Cement	None
Pigments for Cement and Concrete	BS 1014
Pulverized-Fuel Ash (PFA) for Use in Concrete	BS 3892
Masonry Cement	BS 5224

AGGREGATES	
Aggregate from Natural Sources for Concrete (Including Granolithic)	BS 882 and 1201
Lightweight Aggregates for Concrete (Expanded Clay, Shale, Perlite, Pumice Pulverized Fuel Ash and Exfoliated Vermiculite)	BS 3797
Foamed or Expanded Blastfurnace Slag Lightweight Aggregate for Concrete	BS 877
Air-Cooled Blastfurnace Slag Coarse Aggregate for Concrete	BS 1047

Clinker Aggregate for Concrete	BS 1165
Building Sands from Natural Sources	BS 1198–1200

TESTING CONCRETE

Methods of Testing Concrete	BS 1881
Methods for Sampling and Testing of Mineral Aggregates, Sands and Fillers	BS 812
Methods of Sampling and Testing Lightweight Aggregates for Concrete	BS 3681
Tests for Water for Making Concrete	BS 3148
Test Sieves	BS 410
Ready-Mixed Concrete	BS 1926

STEEL FOR REINFORCEMENT AND OTHER USES

Expanded Metal (Steel) for General Purposes	BS 405
Hard Drawn Mild Steel Wire for the Reinforcement of Concrete	BS 4482
Hot Rolled Steel Bars for the Reinforcement of Concrete	BS 4449
Cold Worked Steel Bars for the Reinforcement of Concrete	BS 4461
Steel Fabric for the Reinforcement of Concrete	BS 4483
Steel Wire for Prestressed Concrete	BS 2691
Stress-Relieved Wire Steel Strand for Prestressed Concrete	BS 3617
Galvanized Steel Reinforcement	None
Structural Steel Sections	BS 4
Galvanized Iron and Steel Wire for Telegraph and Telephone Purposes	BS 182–4
Galvanized Wire Netting	BS 1485
The Use of Structural Steel in Building	BS 449
Stainless Steel Bars	BS 970
Stainless Heat Resisting Plate, Sheet and Strip	BS 1449: Part 4

PRECAST AND PREFORMED MATERIALS

Precast Concrete Flags	BS 368
Concrete Flooring Tiles and Fittings (Dimensions and Workmanship Only)	BS 1197
Terrazzo Tiles	BS 4131

Asphalt Tiles for Paving and Flooring (Natural Rock Asphalt)	BS 1324
Acid-Resisting Bricks and Tiles	BS 3679
Clay Tiles for Flooring	BS 1286
Asbestos-Cement Tiles for Promenade Use	None

JOINT SEALANTS

Hot Applied Joint Sealing Compounds for Concrete Pavements	BS 2499
Corrugated Copper Jointing Strip for Expansion Joints for Use in General Building Construction	BS 1878
Two-Part Polysulphide-Based Sealing Compounds for the Building Industry	BS 4254
Mastic asphalt for Tanking and Damp-Proof Courses	BS 1097 and 1418
Mastic Asphalt for Roofing (Natural Rock Asphalt Aggregate)	BS 988 and 1162
Water Bars, Rubber	None
Water Bars, PVC	None
Neoprene Joint Sealants	None
Mastic Asphalt for Flooring	BS 1076, 1410 and 1451
Low and Medium Density Polyethylene Sheet for General Purposes	BS 3012
Materials for Damp-Proof Courses	BS 743
Cold Poured Joint Sealants	BS 5212
One-Part Gun-Grade Polysulphide-Based Sealants	BS 5215

GENERAL

Conversion Factors and Tables	BS 350
Bending Dimensions and Schedules of Bars for the Reinforcement of Concrete	BS 4466
Glossary of Terms for Concrete and Reinforced Concrete	BS 2787
Glossary of General Building Terms	BS 3589
Fire Tests on Building Materials and Structures	BS 476
Building Drawing Practice	BS 1192
Schedule of Weights of Building Materials	BS 648
Methods of Determining Thermal Properties, with Definitions of Thermal Insulating Terms	BS 874

Glossary of Terms of Internal Plastering, External Rendering and Floor Screeds	BS 4049
Recommendations for the Co-ordination of Dimensions in Building. Controlling Dimensions	BS 4330
Code of Basic Data for the Design of Buildings	CP 3
Thermal Insulation in Relation to the Control of Environment	Chapter II
Sound Insulation and Noise Reduction	Chapter III
Precautions Against Fire	Chapter IV
Dead and Imposed Loads and Wind Loads	Chapter V
Engineering and Utility Services	Chapter VII
Heating and Thermal Insulation	Chapter VIII
Demolition	CP 94
Metal Scaffolding	CP 97: Parts 1–3
Foundations and Sub-structures for Non-Industrial Buildings of Not More Than Four Storeys	CP 101
Protection of Buildings Against Water from the Ground	CP 102
The Structural Use of Concrete	CP 110: Parts 1 and 2
Structural Recommendations for Load-Bearing Walls	CP 111
Structural Use of Reinforced Concrete in Buildings	CP 114
The Structural Use of Prestressed Concrete in Buildings	CP 115
The Structural Use of Precast Concrete	CP 116
Composite Construction in Structural Steel and Concrete	CP 117
The Protection of Structures Against Lightning	CP 326
Site Investigations	CP 2001
Earthworks	CP 2003
Foundations	CP 2004
Sewerage	CP 2005
Lining Vessels and Equipment for Chemical Purposes	CP 3003
Cathodic Protection	CP 1021
Safety Precautions in the Construction of Large Diameter Boreholes for Piling and Other Purposes	BS 5573
Pipelines	CP 2010: Parts 1–5
Noise Control on Construction Sites	BS 5228

The Structural Use of Concrete for Retaining Aqueous Liquids	BS 5337
Protection of Iron and Steel Structures from Corrosion	BS 5493
Sheet Roof and Wall Coverings	BS 5247
Methods for Specifying Concrete	BS 5328
Wall Tiling. Part 1—Internal, Part 2—External	BS 5385
Stone Masonry Code of Practice	BS 5390
Methods of Test for Metallic and Related Coatings	BS 5411
Surface Finish of Blast-Cleaned Steel for Painting	BS 4232
Anodic Oxidation Coatings on Aluminium	BS 1615
Sprayed Metal Coatings	BS 2569
Sherardized Coatings on Iron and Steel	BS 4921
Methods of Testing Cement	BS 4550
Code of Practice for Preservation of Timber	BS 5589
Code of Practice for the Safe Use of Explosives in the Construction Industry	BS 5607
The Structural Use of Masonry. Part 1—Unreinforced Masonry	BS 5628
Code of Practice for External Rendered Finishes	BS 5262 (Formerly CP 221)
Code of Practice for Internal Plastering	BS 5492 (Formerly CP 211)
Code of Practice for Accuracy in Building	BS 5606
Commentary on Corrosion at Bi-metallic Contacts and its Alleviation	PD 6484

*A*1.2. *United States of America*

American Society for Testing and Materials

Specification for Malleable Iron Castings	A 47-68
Specification for Zinc (Hot-Galvanized) Coatings on Products Fabricated from Rolled, Pressed, and Forged Steel Shapes, Plates, Bars and Strip	A 123-73
Specification for Welded Steel Wire Fabric for Concrete Reinforcement	A 185-73
Specification for Uncoated Seven-Wire Stress-Relieved Strand for Prestressed Concrete	A 416-68
Specification for Welded Deformed Steel Wire Fabric for Concrete Reinforcement	A 497-72

Specification for Deformed and Plain Billet-Steel Bars for Concrete Reinforcement	A 615-72
Specification for Rail-Steel Deformed and Plain Bars for Concrete Reinforcement	A 616–72
Specification for Axle-Steel Deformed and Plain Bars for Concrete Reinforcement	A 617-72
Method of Making and Curing Compressive and Flexural Strength Test Specimens in the Field	C 31-69
Specification for Concrete Aggregates	C 33-74
Method of Test for Compression Strength of Cylindrical Specimens	C 39-72
Specification for Concrete Building Bricks	C 55-71
Specification for Solid Load-Bearing Concrete Masonry Units	C 145-71
Specification for Portland Cement	C 150-74
Method of Sampling Fresh Concrete	C 172-71
Specification for Air-Entraining Admixtures for Concrete	C 260-73
Specification for Liquid Membrane-Forming Compounds for Curing Concrete	C 309–74
Specification for Lightweight Aggregates for Structural Concrete	C 330-69
Method of Test for Drying Shrinkage of Concrete Block	C 426-70
Specification for Aggregates for Masonry Grout	404-70
Test for Abrasion Resistance of Concrete	418-68
Tests for Resistance of Concrete to Rapid Freezing and Thawing	C 666-73 and C 671
Specification for Preformed Expansion Joint Filler for Concrete (Bituminous Type)	D 994-71
Specification for Concrete Joint Sealer, Hot-Poured Elastic Type	1190-64 (1970)
Testing Concrete Joint Sealers	1191-64 (1970)
Specification for Concrete Joint Sealer, Cold-Application Type	1850-67 (1972)
Testing Concrete Joint Sealers, Cold-Application Type	1851-67 (1972)
Specification for Deformed Steel Wire for Concrete Reinforcement	A 496-72
Specification for Cold-Drawn Steel Wire for Concrete Reinforcement	A 82-72

Specification for Fabricated Steel Bar or Rod Mats for Concrete Reinforcement	A 184-65 (1972)
Specification for Uncoated Stress-Relieved Wire for Prestressed Concrete	A 421-65 (1972)
Descriptive Nomenclature of Constituents of Natural Mineral Aggregates	C 294
Specification for Sheet Materials for Curing Concrete	C 171
Test for Resistance to Abrasion of Small Size Coarse Aggregate by Use of the Los Angeles Machine	C 131-76
Test for Resistance to Abrasion of Large Size Coarse Aggregate by Use of the Los Angeles Machine	C 535-69
Test for Scaling Resistance of Concrete Surfaces Exposed to Deicing Chemicals	C 672-76
Recommended Practice for Evaluation of Frost Resistance of Coarse Aggregates in Air-entrained Concrete by Critical Dilation Procedures	C 682-75
Test for Effect of Organic Impurities in Fine Aggregate on Strength of Mortar	C 87-69
Definitions of Terms Relating to Concrete and Concrete Aggregates	C 125-76
Test for Potential Reactivity of Aggregates (Chemical Method)	C 289-71
Test for Potential Alkali Reactivity of Carbonate Rocks for Concrete Aggregates (Rock Cylinder Method)	C 586-69
Test for Specific Gravity and Absorption of Coarse Aggregate	C 127-77
Test for Specific Gravity and Absorption of Fine Aggregate	C 128-73
Test for Total Moisture Content of Aggregate by Drying	C 566-67
Specification for Blended Hydraulic Cements	C 595-76
Test for Unit Weight, Yield, and Air Content (Gravimetric) of Concrete	C 138-77
Test for Air Content of Hydraulic Cement Mortar	C 185-75
Test for Ball Penetration in Fresh Portland Cement Concrete	C 360-63
Specification for Chemical Admixtures for Concrete	C 494-71

Specification for Fly Ash and Raw or Calcined Natural Pozzolans for Use in Portland Cement Concrete	C 618-73
Method of Making, Accelerated Curing, and Testing of Concrete Compression Test Specimens	C 684-74

American Welding Society

Recommended Practices for Welding Reinforcing Steel, Metal Inserts and Connections in Reinforced Concrete Construction	AWS D12. 1-61

American Concrete Institute

Specification for Structural Concrete for Buildings	ACI 301-72 (Revised 1973)
Recommended Practice for Measuring, Mixing, Transporting, and Placing Concrete	ACI 304-73
Recommended Practice for Hot Weather Concreting	ACI 305-72
Recommended Practice for Cold Weather Concreting	ACI 306-66 (Reaffirmed 1972)
Recommended Practice for Curing Concrete	ACI 308-71
Recommended Practice for Consolidation of Concrete	ACI 309-72
Building Code Requirements for Reinforced Concrete	ACI 318–77
Recommended Practice for Concrete Formwork	ACI 347-68
Guide for the Protection of Concrete Against Chemical Attack by Means of Coatings and Other Corrosion Resistant Materials	ACI Committee 515 Report (63-59)
Recommended Practice for Design and Construction of Concrete Bins, Silos, and Bunkers for Storing Granular Materials; and Commentary 1977	ACI 313-77
Structural Plain Concrete	ACI 322.1-322.9
Concrete Shell Structures; Practice and Commentary	ACI 334.1-18

Recommended Practice for Shotcrete	ACI 506-66
Recommended Practice for Selecting Proportions for Normal and Heavyweight Concrete	ACI 211.1-74
Recommended Practice for Evaluation of Compression Test Results of Field Concrete	ACI 214

*A*1.3. *Canada*

Concrete Materials and Methods of Concrete Construction	CSA A23.1-M
Portland Cements (Five Types)	CSA A5-M
Masonry Cement	CSA A8-M
Sampling Aggregates for Use in Concrete	CSA A23.2-M1A
Organic Impurities in Sand for Concrete	CSA A23.2-M7A
Measuring Mortar Strength Properties of Five Aggregates	CSA A23.2-M8A
Tests for Impurities in Aggregate	CSA A23.2-M4A
Clay Lumps in Concrete	CSA A23.2-M3A
Sieve Analysis of Fine and Coarse Aggregate	CSA A23.2-M2A
Methods of Test for Moisture in Aggregate	CSA A23.2-M6A
Methods of Sampling Plastic Concrete	CSA A23.2-M1C
Slump of Concrete	CSA A23.2-M5C
Methods of Test for Air Content of Fresh Concrete	CSA A23.2-M4C
Methods of Test for Density of Concrete	CSA A23.2-M6C
Making and Curing Concrete Compression and Flexure Test Specimens in the Field	CSA A23.2-M3C
Splitting Tensile Strength of Cylindrical Concrete Specimens	CSA A23.2-M13C
Accelerating the Cure of Concrete Cylinders and Determining Their Compressive Strength	CSA A23.2-M10C
Obtaining and Testing Drilled Cores for Compressive Strength Testing	CSA A23.2-M14C
Air Content of Plastic Concrete	CSA A23.2-M7C
Compressive Strength of Cylindrical Concrete Specimens	CSA A23.2-M9C
Test for Flexure Strength of Concrete	CSA A23.2-M8C
Abrasion Tests for Concrete	CSA A23.2-M16A & M17A

Aggregates, Tests for Strength and Durability	CSA A23.2-M9A
Aggregates, Tests for Strength and Durability	CSA A23.2-M14A
Aggregates, Tests for Strength and Durability	CSA A23.2-M15A
Aggregates, Particle Shape Tests	CSA A23.2-M13A
Aggregates, Tests for Bulk and Mass Density	CSA A23.2-M10A
Aggregates, Tests for Relative Density	CSA A23.2-M12A
Aggregates, Tests for Absorption and Moisture Content	CSA A23.2-M11A
Specifications, Methods of Testing Air-Entraining Admixtures	CSA A266.1
Specifications and Tests for Accelerating Admixtures	CSA A266.2
Specifications and Tests for Pozzolanic Admixtures	CSA A266.3
Specifications and Tests for Retarding Admixtures	CSA A266.4
Compounds, Concrete Curing, Liquid, Membrane Forming	CSA 90-GP-1a
Standard for Guide to Selection of Sealants on a Use Basis	CSA 19-GP-23
Sealing Compounds—Two-part Polysulphide	CSA 19-GP-3b
Sealing Compounds—One-component Silicone Based	CSA 19-GP-5b
Sealing Compounds—One-component	CSA 19-GP-13b

*A*1.4. *Australia*

For a complete list of Publications and Subject Index of Australian Standards, 1976, reference should be made to the Standards Association of Australia, Standards House, 80–86 Arther Street, North Sydney, NSW.

The following are some of the principal standards relating to cement and concrete:

Portland Cement	AS 2
Masonry Cement	AS 1316
Blended Cement	AS 1317
Methods of Testing Concrete	AS 100 *et seq.*
Concrete Structures Code	AS 1480
Prestressed Concrete Code	AS 1481
Manual of Physical Testing of Portland Cement	AS MP27
Admixtures	A173, CA 58, MP 20, 1478–9

Aggregates for Concrete	1465–7
Tests on Aggregates	1141
Readymixed Concrete	A64, 1379
Reinforcement for concrete, bars, wire and rods	A81–4, A92, A97, 1302–4

*A*1.5. *New Zealand*

AGGREGATES	
Methods of Test for Water and Aggregates for Concrete	NZ 3111: 1974
Water and Aggregate for Concrete	NZS 3121: 1974
Lightweight Aggregate for Structural Concrete	NZS 1958: 1965
Lightweight Aggregate for Concrete Masonry Units	NZS 1959: 1965
Lightweight Aggregates for Insulating Concrete	NZS 1960: 1965
CEMENTS	
Portland Cement (Ordinary, Rapid Hardening, and Modified)	NZS 3122: 1974
Portland Pozzolan Cement	NZS 3123: 1974
CONCRETE CONSTRUCTION	
Reinforced and Plain Concrete Construction	NZS 3101P: 1970
CONCRETE MIXERS	
Concrete Mixers	NZS 3105: 1975
CONCRETE TESTING	
Methods of Test for Concrete	NZS 3112: 1974
FIRE RESISTANCE	
Reports on Fire Resistance Ratings of Elements of Building Structure	NZ MP9
Fire Resistance Ratings of Walls and Partitions: Structures of Concrete Masonry Blocks	NZ MP9/1: 1962
Fire Resistance Ratings of Walls and Partitions: Structural Concrete	NZ MP9/2: 1962
Fire Resistance of Prestressed Concrete	NZ MP9/5: 1972
Fire Resistance of Beams and Columns, Excluding Timber	NZ MP9/6: 1966

Fire Resistance Ratings of Floor/Ceiling Combinations	NZ MP9/7: 1966
Foundations	
Metric Handbook to NZS 4204P: 1973 Code of Practice for Foundations for Buildings Not Requiring Specific Design	NZ MP420400: 1973
Metric Handbook to NZS 4205P: 1973 Code of Practice for Design of Foundations for Buildings	NZ MP420500: 1973
Foundations	NZS 4204P: 1973
Foundations	NZS 4205P: 1973
Lightweight Concrete (See also Aggregates)	
Manufacture and Use of Structural and Insulating Lightweight Concrete	NZ 3152: 1974
Precast Lightweight Concrete Panels and Slabs	NZS 3141P: 1970
Masonry (See also NZS 1900 below)	
Concrete Masonry Units	NZS 3102P: 1974
Pavements	
Cold Poured Joint Sealant for Concrete Pavements (Endorsed by NZ)	BS 5215: 1975
Precast Concrete	
Precast Lightweight Concrete Panels and Slabs	NZ 3151: 1974
Prestressed Concrete	
Prestressed Concrete Amendment No. 1: 1970	NZSR 3R32: 1968
Steel Wire for Prestressed Concrete (Amended to suit NZ conditions from BS 2691: 1969)	NZ 1417: 1971
Metric Handbook to NZSR 32: 1968 Prestressed Concrete	NZ MP32x000: 1974
Ready Mixed Concrete	
Ready Mixed Concrete Production	NZS 2086: 1974
Metric Handbook to NZS 2086: 1968 Ready Mixed Concrete	NZ MP208600: 1975

REINFORCED CONCRETE

Code of Practice for Reinforced Concrete—Design	NZ 3101 P: 1970
Metric Handbook to NZS 3101 P: 1970 Code of Practice for Reinforced Concrete Design	NZ MP310100: 1973

REINFORCEMENT

Hot Rolled Steel Bars for Concrete Reinforcement	NZS 3402P: 1973
Hot-Dip Galvanised Corrugated or Profiled Steel Sheet	NZS 3403
Hard Drawn Mild Steel Wire for Concrete Reinforcement	NZS 3421: 1975
Hot-Dip Galvanised Plain Steel Sheet and Strip (Metricated 1976)	NZS 3441
Welded Fabric of Drawn Steel Wire for Concrete Reinforcement	NZS 3422: 1975

SANDS

Sands for Mortars, Plasters and External Renderings Amendment No. 1: 1971 (Revised 1976)	NZ 2129

SHELLS

The Design and Construction of Shell Roofs	NZ 1826: 1964

SWIMMING POOLS

Code of Practice for Swimming Pools	NZS 4441: 1972

NZS 1900 (MODEL BUILDING BYLAW)

Masonry Amendment No. 1: 1973 (Revised 1976)	Ch 6.2: 1964
Design and Construction Amendment No. 1: 1965, No. 2: 1973 (Revised 1976) (to be Reconstituted Incorporating NZS 2086)	Ch 9.2: 1964
Concrete—General Requirements and Materials and Workmanship Amendments No. 1: 1970, No. 2: 1971, No. 3: 1973 (Revised 1976) (to be Reconstituted Incorporating NZS 2086)	Ch 9.3A: 1970
Special Structures	Ch 11
Concrete Structures for the Storage of Liquids Amendments No. 1: 1964, No. 2: 1973 (Revised 1976)	11.1: 1964

Metric Handbook to NZS 1900 Chapter 9.3A: 1970 Concrete Design and Construction General Requirements — MP 190093A:

*A*1.6. *India*

Code of Practice for Plain and Reinforced Concrete	IS 476
Portland Cement: Ordinary, Rapid Hardening and Low Heat	IS 269
Portland Blastfurnace Cement	IS 455
Pozzolanic Cement	IS 1489
Methods of Test for Strength of Concrete	IS 516
Masonry Cement	IS 3466
High Alumina Cement for Structural Use	IS 6452

*A*1.7. *South Africa*

STANDARDS (METRIC UNITS, UNLESS OTHERWISE INDICATED)

Metal Ties for Cavity Walls	28-1972
Bending Dimensions of Bars for Concrete Reinforcement	82-1976
Test Sieves	197-1971
Bituminous Damp-Proof Courses	248-1973
Portland Cement and Rapid-Hardening Portland Cement	471-1971
Limes for Use in Building	523-1972
Concrete Building Blocks	527-1972
Precast Concrete Paving Slabs	541-1971
Portland Blastfurnace Cement	626-1971
Aggregates of Low Density	794-1973
Portland Cement 15 and Rapid-Hardening Cement 15	831-1971
Ready Mixed Concrete	878-1970
Steel Bars for Concrete Reinforcement	920-1969
Precast Concrete Kerbs and Channels (*not metric*)	927-1969
Standard Form of Specification for Concrete Work	973-1970
Prestressed Concrete Pipes	975-1970
Precast Reinforced Concrete Culverts	986-1970
Cement Bricks	987-1970

Modular Co-ordination in Building	993-1972
Welded Steel Fabric for Concrete Reinforcement	1024-1974
Aggregates from Natural Sources	1083-1976

CODES OF PRACTICE

Waterproofing of Buildings	021-1973
Sewer and Drain Jointing (*not metric*)	058-1955
Safe Application of Masonry-type Facings to Buildings	073-1974
Pile Foundations	088-1972
The Structural Design and Installation of Precast Concrete Pipelines	0102-1968
Floor Finishes on Concrete	0109-1969

STANDARD BUILDING REGULATIONS

Note: The Standard Building Regulations cover all aspects of building construction in areas controlled by local authorities, although not all cities and towns in South Africa have adopted them. The main chapter headings cover definitions, administration in the four provinces of South Africa, loads and forces, foundations, plain and reinforced concrete, structural steelwork, structural timber, masonry and walling, miscellaneous materials and construction, water supply, lighting, drainage and sewerage, ventilation, fire protection, public safety, urban aesthetics and advertising.

DEPARTMENT OF INDUSTRIES

No. R.1830.] [23rd October, 1970.

STANDARDS ACT, 1962

STANDARD BUILDING REGULATIONS

In terms of section 14*bis* (1) of the Standards Act, 1962 (Act No. 33 of

1962) the Council of the South African Bureau of Standards, with the approval of the Minister of Economic Affairs, hereby publishes standard building regulations under the chapter headings listed in this notice.

LIST OF CHAPTERS

Chapter

1 DEFINITIONS AND INTERPRETATION
2 ADMINISTRATION, PART I, FOR THE PROVINCE OF THE CAPE OF GOOD HOPE
ADMINISTRATION, PART II, FOR THE PROVINCE OF THE TRANSVAAL
ADMINISTRATION, PART III, FOR THE PROVINCE OF NATAL
ADMINISTRATION, PART IV, FOR THE PROVINCE OF THE ORANGE FREE STATE
3 LOADS AND FORCES
4 FOUNDATIONS
5 PLAIN AND REINFORCED CONCRETE
6 STRUCTURAL STEELWORK
7 STRUCTURAL TIMBER
8 MASONRY AND WALLING
9 MISCELLANEOUS MATERIALS AND CONSTRUCTION
10 WATER SUPPLY
11 LIGHTING
12 DRAINAGE AND SEWERAGE
13 VENTILATION
14 FIRE PROTECTION
15 PUBLIC SAFETY
16 URBAN AESTHETICS
17 ADVERTISING

A1.8. Federal Republic of Germany

Concrete for Sewerage Units, Requirements for Manufacture and Testing	DIN. 4281
Refers to DIN. 1045, 0147, 0148, 4225, 4030	
Precast Reinforced Concrete	4255
Evaluation of Liquids, Soils and Gases Aggressive to Concrete	4030
Concrete and Reinforced Concrete Structures—Design and Construction	1045
Code of Practice for Structural Use of Concrete	1047
Design Loads for Buildings	1055

Water Retaining Structures—Codes of Practice for Design, Construction and Operation. Part 1—Dams	19700
Reinforcing Steel	488
Methods of Testing Concrete	1048
Portland Cement	1164
Water Pressure, Retaining Bituminous Seals for Structures. Code of Practice for Design and Construction	4031
Sealing of Structures Against Ground Moisture. Codes of Practice for Construction	4117
Aggregates for Concrete	4226
Symbols for Structural Design Calculations in Civil and Structural Engineering	1080
Quality Control in Concrete and Reinforced Concrete Construction	1084
Portland and Other Cements	1164
Contract Procedure for Building Work. Part C—Technical Specification for Grouting	18309
Contract Procedure for Building Work, General, Technical Specification for Damp-Proofing against Water under Pressure	18336
Concrete for Sewerage Works, Manufacture and Testing	4281

*A*1.9. *Austria*

Building Contracts, Placing Contracts	A 2050
Building Construction, General Conditions for Building Contracts	ONORM B 2110
Concrete and Reinforced Concrete Construction	ONORM B 2211
Testing Concrete	ONORM B 3303
Water, Soils and Gases Aggressive to Concrete: Evaluation and Chemical Analysis	ONORM B 3305
Concrete Structures, Design and Construction	ONORM B 4200
Prestressed Concrete Structures	ONORM B 4250
Portland Cements, Types OCI, OCII, HSC, and Blastfurnace Cements	ONORM B 3310
Steel Reinforcement for Concrete	ONORM B 4200: Part 7

A1.10. *Netherlands*

Regulations for Concrete (these partly supersede NEN 1009)	NEN 3861 and 3862
Steel Reinforcement for Concrete	NEN 6008
Portland Cement	N 481
Portland Blastfurnace Cement	N 483 and 484
Pozzolanic Cement	N 618

A1.11. *Italy*

Portland Cement, Blastfurnace Cement and Pozzolanic Cements	UNI 595
Testing of Concrete	UNI 6132 and 6135

A1.12. *Belgium*

Concrete Construction	NBN 15
201–205, 211–214, 216, 218, 220, 227 and 250 Concrete Tests	NBN 15
Steel Reinforcement for Concrete	NBN A24.301–303
Portland Cement	NBN 48
Portland Blastfurnace Cement	NBN 130, 131 and 198

A1.13. *France*

Reinforced Concrete Code	CC BA 68
Structural Steelwork Code	CM 66
Standard Contracts for Private Building Work, General Clauses	PO 3–011
Portland Cement	NF P 15–302
Portland Blastfurnace Cement	303, 304, 305 and 311
Tests for Concrete	NF P 18–102, 404, 405, 406
Steel Reinforcement for Concrete	NF A.35.015 and 016.

APPENDIX 2

CONVERSION FACTORS AND CONSTANTS

$1\,m^2 = 10{\cdot}7\,ft^2$
$1\,ft^2 = 0{\cdot}093\,m^2$
$1\,kg = 2{\cdot}26\,lb$
$1\,lb = 0{\cdot}45\,kg$
$1\,m = 3{\cdot}28\,ft$
$1\,ft = 0{\cdot}31\,m$
$1\,lbf/ft^2 = 4{\cdot}85\,kg/m^2$

CONVERSION FROM IMPERIAL TO METRIC

D_i = density in lb/ft^3
D_m = density in kg/m^3

$$D_m = \frac{D_i}{62{\cdot}4} \times \frac{1000}{1} = 16\,D_i \text{ (approx)}$$

DENSITY OF STRUCTURAL CONCRETE MADE WITH AGGREGATES FROM NATURAL SOURCES

$$148\,lb/ft^3 = 4000\,lb/yd^3 = 2385\,kg/m^3 = 2400\,kg/m^3 \text{ (approx)}$$

BULK DENSITIES OF CONCRETING MATERIALS

cement	$1400\,kg/m^3 = 88\,lb/ft^3$
sand	$1600\,kg/m^3 = 100\,lb/ft^3$
coarse aggregate	$1450\,kg/m^3 = 91\,lb/ft^3$

(These figures are very approximate.)

FORCE

$1\,lbf$ =	$4{\cdot}45\,N$
$1000\,lbf/in^2$ =	$7\,N/mm^2$ (approx)

143 lbf/in^2 — 1 N/mm^2 (approx)

1 kN/mm^2 = 1000 N/mm^2

1 MN/m^2 = 10^6 N/m^2 = 1 N/mm^2

THERMAL TRANSMITTANCE (U) VALUE

$$Btu/ft^2\,h\,°F = \frac{W/m^2\,°C}{5{\cdot}678}$$

$$= 0{\cdot}176(W/m^2\,°C$$

$$W/m^2\,°C = 5{\cdot}678\,Btu/ft^2\,h\,°F$$

THERMAL CONDUCTIVITY, COEFFICIENT OF HEAT TRANSFER

1 Btu ft/ft^2 h °F = 1·731 W/m °C.

COEFFICIENT OF THERMAL EXPANSION OF CONCRETE

Imperial: Range: $3{\cdot}5 \times 10^{-6}$ to $6{\cdot}5 \times 10^{-6}$ °F

Metric: Range: $6{\cdot}3 \times 10^{-6}$ to $11{\cdot}7 \times 10^{-6}$ °C

To convert lb/yd^3 of compacted concrete to kg/m^3 multiply by 0·6.
To convert kg/m^3 of compacted concrete to lb/yd^3 multiply by 1·67.

Concrete strengths are designated in newtons per square millimetre (N/mm^2). Sometimes mega-newtons per square metre (MN/m^2) are used. The two units are equal numerically.

MODULUS OF ELASTICITY OF CONCRETE

$$E = 4{\cdot}5 \times 10^6$$

$$\text{Range: } E = 2{\cdot}5 \times 10^2 \text{ to } 6 \times 10^2$$

SPECIFIC GRAVITY

	Imperial	Metric
water	1	1
cement, Portland	3·12	3·12
cement, high alumina	3·20	3·20
pit sand	2·65	2·65
flint gravel	2·55	2·55
limestone	2·80	2·80
granite	2·75	2·75
basalt (Whinstone)	2·90	2·90
concrete structural	2·38	2·38

Specific Heat

concrete	0·25 Btu	0·063 kcal
mild steel	0·12 Btu	0·031 kcal

Latent Heat

fusion of ice	144 Btu/lb	37 kcal/g
evaporation of water	970 Btu/lb	250 kcal/g

1 Btu/lb = 2326 J/kg

APPENDIX 3

CONSTRUCTION ASSOCIATIONS AND RESEARCH ORGANISATIONS

*A*3.1. *United Kingdom*

The Cement and Concrete Association
52 Grosvenor Gardens
London SW1W 0AQ
Telephone: 01-235 6661

The Building Research Establishment
Garston
Watford, Herts
Telephone: 09273 76612

The British Standards Institution
2 Park Street
London W1A 2BS
Telephone: 01-629 9000

The Building Centre
26 Store Street
London WC1
Telephone: 01-636 5400

The British Concrete Pumping Association
7/10 Stray Court
Princes Villa Road
Harrogate HG1 5RJ
Telephone: 0423 57020

British Precast Concrete Federation
60 Charles Street
Leicester LE1 1FB
Telephone: 0533 28627

British Ready-Mixed Concrete Association
Shepperton House
Green Lane
Shepperton, Middlesex TW17 8DN
Telephone: 98 43232

The Aluminium Federation
60 Calthorpe Road
Fiveways
Birmingham 15
Telephone: 021-454 3805

The Brick Development Association
19 Grafton Street
London W1X 3LE
Telephone: 01-409 1021

British Ceramic Tile Council
Federation House
Stoke-on-Trent, Staffs
Telephone: 0782 45147

Corrosion Advice Bureau
British Steel Corporation
Shotwick, Near Chester
Telephone: 0244 812345

The Clay Pipe Development Association
30 Gordon Street
London WC1
Telephone: 01-388 0025

The Construction Industry Research and Information Association
6 Storeys Gate
London SW1
Telephone: 01-839 6881

The Construction Steel Research and Development Organisation (Constrado)
12 Addiscombe Road
Croydon CR9 3JH
Telephone: 01-686 0366

The Mastic Asphalt Council
24 Grosvenor Gardens
London SW1W 0DH
Telephone: 01-730 7175

Reinforcement Manufacturers' Association
16 Tooks Court
London EC4
Telephone: 01-242 4259

The Stainless Steel Development Association
British Steel Corporation
PO Box 150
Sheepcote Lane
Sheffield S9 1TR
Telephone: 0742 40311

The Steel Sheet Information and Development Association
Albany House
Petty France
London SW1
Telephone: 01-799 1616

Copper Development Association
Orchard House
Mutton Lane
Potters Bar, Herts EN6 3AP
Telephone: 77 50711

Zinc Development Association
34 Berkeley Square
London W1X 6AJ
Telephone: 01-499 6636

The Association of Gunite Contractors
24 Ormond Road
Richmond, Surrey TW1O 6TH
Telephone: 01-948 4151

*A*3.2. *West Germany*

Bundesverband der Deutschen Zementindustrie e.V. (Cement industry and technical advisory services)
Riehlerstrasse 8
(Postfach 140105)
5 *Köln* 1
T 73 00 76 C Zementverband
Tx 08881603

Forschungsinstitut der Zementindustrie (Research Institute of the cement industry)
Tannenstrasse 2
4 *Düsseldorf-Nord*
T 43 44 51 C Zementforschung
Tx 08584876

Verein Deutscher Zementwerke e.V. (Technical association of the cement industry)
Tannenstrasse 2
4 *Düsseldorf-Nord*
T 43 44 51 C Zementforschung
Tx 08584876

Deutscher Beton-Verein e.V. (Concrete)
Bahnhofstr 61
62 *Wiesbaden*

*A*3.3. *United States of America*

Portland Cement Association (Research, information and technical advisory services)
Old Orchard Road
Skokie, Illinois
T (312) 9666200

American Concrete Institute
P.O. Box 4754
Redford Station
Detroit, Michigan 48219

(Technical and Educational Committee activities and publications for manufacturers and users of concrete)

Architectural Precast Association (APA)
2201 East 46th Street
Indianapolis, Indiana 46205
T (317) 2511214

(Precast concrete)

National Concrete Masonry Association (NCMA)
Pompino East Building—1800 Kent Street
P.O. Box 9185
Rosslyn Station
Arlington, Virginia 22209
T (703) 5240815

(Precast concrete)

National Precast Concrete Association (NPCA)
2201 East 46th Street
Indianapolis, Indiana 46205
T (317) 2530486

(Precast concrete)

Prestressed Concrete Institute (PCI)
20 North Wacker Drive
Chicago, Illinois 60606
T (312) 3464071

(Prestressed concrete)

American Concrete Pipe Association (ACPA)
1501 Wilson Boulevard
Arlington, Virginia 22209
T (703) 5243939

(Concrete pipes)

American Concrete Pressure Pipe Association (ACPPA)
1501 Wilson Boulevard
Arlington, Virginia 22209
T (703) 5243939

(Concrete pipes)

Concrete Reinforcing Steel Institute (CRSI) 228 North LaSalle Street *Chicago*, Illinois 60601 T (312) 3725059	(Concrete reinforcing steel)
National Ready Mixed Concrete Association (NRMCA) 900 Spring Street *Silver Spring*, Maryland 20910 T (310) 5871400	(Ready mixed concrete)
Expanded Shale, Clay and Slate Institute 1041 National Press Building *Washington*, D.C. 20004	(Structural lightweight concrete)

*A*3.4. *Canada*

Canadian Portland Cement Association 160 Bloor St. East *Toronto*, Ontario M4W 19B T (416) 920	(Research, information and technical advisory services)
Canadian Prestressed Concrete Institute (CPCI) 120 Eglington Avenue, East *Toronto* 12, Ontario T (416) 4895616	(Prestressed concrete)
National Concrete Producers' Association (NCPA) 3500 Dufferin Street Suite 101 *Downsview* 460, Ontario T (416) 6301204	(Concrete industry)

A3.5. Australia

Cement and Concrete Association of Australia
147–152 Walker Street
North Sydney, NSW 2060

Concrete Institute of Australia
147 Walker Street
North Sydney, NSW 2060
T 92 0316
(Concrete research)

Concrete Masonry Association of Australia
147 Walker Street
North Sydney, NSW 2060
T 92 0316
(Concrete masonry)

National Ready Mixed Concrete Association
332 Albert Street
Melbourne, Victoria 3001
T 419 1313
(Ready mixed concrete)

Precast Concrete Manufacturers Association
12 O'Connell Street
Sydney, 2001
T 250 5401
(Precast concrete)

C.S.I.R.O. Division of Building Research
P.O. Box 56
Highett, Victoria 3190

Department of Construction, Experimental Building Station
P.O. Box 30
Chatswood, NSW 2067

It should be noted that most Universities carry out research on concrete.

*A*3.6. *New Zealand*

NZ Portland Cement Association
Box 2792
Wellington

(Information and technical advisory services)

NZ Concrete Masonry Association
Box 9130
Wellington

(Concrete masonry)

NZ Concrete Products Association
Box 9130
Wellington

(Concrete products)

NZ Prestressed Concrete Institute
11th floor Securities House
126 The Terrace
P.O. Box 969
*Wellington C.*1

(Prestressed concrete)

NZ Ready Mixed Concrete Association
Box 12013
Wellington

(Ready mixed concrete)

NZ CONCRETE RESEARCH AND INFORMATION ORGANISATIONS: ASSOCIATIONS AND RELATED BODIES
Building Research Association of New Zealand
42 Vivian Street
P.O. Box 9375 (Postal)
Wellington

Department of Scientific and Industrial Research
Chemistry Division
Gracefield Road
Lower Hutt, NZ
Private Bag (Postal)
Petone

Department of Scientific and Industrial Research
Physics and Engineering Laboratory
Gracefield Road
Lower Hutt, NZ
Private Bag (Postal)
Lower Hutt

Ministry of Works and Development
Vogel Building
Aitken Street
P.O. Box 12-041 (Postal)
Wellington 1

NZ Concrete Research Association
13 Wall Place
Tawa, NZ
P.O. Box 50-156 (Postal)
Porirua

NZ Institute of Architects
Maritime Building
2–10 Customhouse Quay
P.O. Box 438 (Postal)
Wellington 1

NZ Institution of Engineers
Molesworth House
101 Molesworth Street
P.O. Box 12-241 (Postal)
Wellington 1

University of Auckland
School of Engineering
24 Symonds Street
Private Bag (Postal)
Auckland

University of Canterbury
School of Engineering
Creyke Road
Ilam
Christchurch 4
Private Bag (Postal)
Christchurch

A3.7. Republic of South Africa

South African Cement Producers Association
P.O. Box 61514
Marshalltown
T 8387502 C safcem
(Cement industry)

National Building Research Institute
P.O. Box 395
Pretoria
T 74 6011 C navorsbou
(Research, information and technical advisory services)

Portland Cement Institute

Head Office:
18 Kew Road
Richmond
Johannesburg
2092

Natal Regional Office:
P.O. Box 90
Westville
Natal
3630

Western Cape Regional Office:
Molteno Street
Goodwood
Cape Province
7460

Eastern Cape Regional Office:
P.O. Box 1540
Port Elizabeth
Cape Province
6000

South African Council for Scientific and Industrial Research
P.O. Box 395
Pretoria
Transvaal
0001
National Building Research Institute
National Institute for Transport and Road Research

Research (only) Organisations

South African Railways Research Laboratories
Leyds Street
Johannesburg
2001

Universities

The Department of Civil Engineering at the following Universities:

University of the Witwatersrand
1 Jan Smuts Avenue
Johannesburg
2001

(This university also has a department of Building Science)

University of Cape Town
Private Bag
Rondebosch
Cape Province
7700

University of Natal
King George V Avenue
Durban
4001

Randse Afrikaanse Universiteit
Posbus 524
Johannesburg
2000

University of Pretoria
Brooklyn
Pretoria
0181

University of Stellenbosch
Stellenbosch
Cape Province
7600

APPENDIX 4

TOLERANCES IN CONSTRUCTION

The recommendations which follow apply to *in situ* concrete construction. The tolerances are those which should be achieved by experienced contractors exercising reasonable care and attention.

If the designer wishes the contractor to work to certain tolerances then those should be written into the contract documents. It should be kept in mind that the smaller the tolerances specified, the higher will be the cost of achieving them. Tolerances finer than those recommended should only be required if there is a sound technical reason for doing so. The recent introduction of finishing techniques for concrete floor slabs, such as power floating and grinding, are likely to eliminate expensive toppings except in special cases.

WALLS

(a) Deviation in specified diameter on circular tanks, per 30 m length of diameter — ±25 mm

(b) Deviation from the vertical or specified inclination:
- height not exceeding 3·00 m — ±15 mm
- height exceeding 3·00 m — ±20 mm

(c) Difference in level across joints, i.e. 'the lip' — ±5 mm

(d) Irregularity in surface finish under 3·00 m straight edge. (This is normally applied to the inside of the tank unless there is special reason to apply it to the outside.) — ±10 mm

(e) Deviation in prescribed overall length of the wall:
- length not exceeding 15·00 m — ±30 mm
- length exceeding 15·00 m but not exceeding 30·00 m — ±40 mm

(f) Deviation from prescribed thickness:
- thickness not exceeding 300 mm — ±10 mm
- thickness exceeding 300 mm — ±15 mm

COLUMNS

(a) Deviation from the vertical:

height not exceeding 6·0 m ±15 mm

height exceeding 6·0 m but not exceeding 10·0 m ±20 mm

(b) Deviation in cross-sectional dimensions:

sectional dimensions not exceeding 300 mm ±5 mm

sectional dimensions exceeding 300 mm but not exceeding 600 mm ±10 mm

FLOORS

(a) Deviation from prescribed level between levelling points ±5 mm

(b) Surface finish:†

(i) irregularities under a 3 m straight edge in any direction ±10 mm

(ii) irregularity across joints, i.e. the 'lip' ±5 mm

† *Note:* In certain structures, e.g. settling tanks containing scraping equipment, closer tolerances may be required. In such cases, these tolerances should be inserted in the specification and the attention of the tenderer should be drawn to them.

Some References on Tolerances

British Standards Institution. *Guidance on Dimensional Coordination in Building*. London, 1977, DD 51. (Contains 7 sections: Section 3 entitled 'Detailed design for fit').

British Standards Institution. BS 5606—*Accuracy in Building*, London, 1978.

Fisher, A. E. Tolerances involving reinforcing bars. *J. ACI*, February 1977, pp. 61–70.

Stevens, A. How accurate is building? *BRE News*, No. 36, summer 1976, pp. 11–13. (A survey during the erection process on 200 construction sites in Great Britain).

Vanderwal, M. L. and Walker, H. C. Tolerances for precast concrete structures. *J. Prestressed Concrete Inst.*, July/August 1976, pp. 44–57. (A proposed list).

INDEX

Abrasion of concrete, 154–6
Absorption of aggregates, 101, 135, 314
Accelerators, *see* Admixtures for concrete
Acids, effect on concrete, 138–41, 144
Acrylic resins, 118, 276
Admixtures for concrete, 105–11, 168, 219, 247
Aggregates, 100–4, 159, 219, 250, 287, 314
Aggressive environment, 126–52
Air entrained concrete, 107, 155, 173, 219, 272
Air entraining agents, *see* Admixtures for concrete
Alkali–aggregate reaction, 136, 137
Alkalinity, 147–51
Alkali-resistant glass, 93
Aluminium, 105, 130, 131
Aluminium sulphate, 146
Analysis of concrete, *see* Quality control and testing of concrete

Base slab, 213; *see also* Floors
Bay
  layout, 217, 218, 220–2, 251
  size, 179, 217, 218, 220, 221, 222, 251
Bending moments, *see* Design
Blow holes in concrete, 234, 238
Bolt holes in formwork and walls, 235
Bond
  strength, *see* Design
  stress, *see* Design
Boreholes, 204, 205
Boxing-out for pipework, 228–30
Brine, effect on concrete, 144
BRMCA, 300
Bronze, 131
Building-in for pipework, 228–31
Bund walls, 273, 274
Butyl rubber, 140, 253

Calcium carbonate, 148
Calcium chloride, 107
Calcium formate, 107
Calcium hydroxide, 148
Capillary pores in concrete, 134
Carbonation, 135, 136
Caustic soda, effect on concrete, 145
Cavitation, 155
Cement
  chemically resistant, 99
  comparison of UK and US Portland cements, 99
  high alumina, 97–9, 141, 151
  Portland
    blastfurnace, 140, 178
    low heat, 177
    ordinary and rapid hardening, 94, 140, 168, 219, 246, 250
    sulphate resisting, 95, 140, 151, 154, 177

Cement—*contd.*
Portland—*contd.*
supersulphated, 96, 140
ultra-high early strength, 96
white and coloured, 95
Cement content of concrete, 134, 150, 151, 153, 155, 157, 160, 167, 170, 203, 206, 219, 235, 247, 251, 276, 314
Cement stabilized materials, 214, 215
Characteristic strength of concrete, 159, 219, 250
Chemical attack
concrete, on, *see* Durability of concrete; Aggressive environment
steel, on, *see* Corrosion of steel
Chloride content
aggregates, of, 102, 153
concrete, of, 102
Chlorides and chloride ions, 101, 103, 151–3
Chlorinated polyethylene, 140, 253, 254
Chlorine, 143
Chromates in cement, 130, 154
Codes of Practice, 6–11, 317–34
Coefficient of thermal expansion of concrete, 13, 177, 336
Cold weather concreting, 167–9, 240, 243
Columns, 232–43
Compacting factor, 158
Compaction of concrete, 158, 222–5, 237, 251
Construction and research associations, 338–48
Contraction joints, *see under* Joints in concrete
Conversion factors and constants, 335–7
Copper, 115, 116, 131
Core testing, 305, 306; *see also* Quality control and testing of concrete
Corner details, 10, 11
Corrosion
inhibitors, 128
steel, of, 127–30, 261–9
Cover meters, 303, 304
Cover to reinforcement, 220, 234, 251, 264, 267, 278
Crack inducers, 178–85, 191, 222
Crack injection, 174, 183, 185
Crack width calculations, 35–9
Cracks in concrete, 169, 171–87, 233, 240, 244, 265, 270, 312, 313
measurement of, 183
Creosote, effect on concrete, 144
Cube strength, 134
Cube testing, 298–300; *see also* Quality control and testing of concrete
Curing concrete, 168, 173, 175, 225–7, 240–3, 247, 251, 265, 279
Cutting concrete, 197, 198

Daywork joints, *see under* Joints in concrete
Degrees of exposure, 18, 19, 202, 203, 311, 312
Density of concrete, 335
Design
circular prestressed tank, 68–89
concrete mixes, of, *see* Mix design
general, 3–17
small reinforced tank, 18–46
swimming pool, 47–67
Dissimilar metals in contact, 130
Drainage of site, 208–10
Drying shrinkage, 185, 240, 276
Ducts for post-tensioning, 269
Durability of concrete, 131–7, 261–9

Elastic design, 6–9
EPDM, 113
Epoxide resins, 118–19, 140, 154, 156, 253, 254, 265, 270, 290, 291
Evaporation from water surface, *see* Water test

Ferro-cement, 94, 123
Ferrous metals, 127–30
Ferrous sulphate, 146
Fibres in concrete, 93, 122, 123
Finishes to concrete, 238–40, 281–7
Finishing floor slabs, 222–5
Flame texturing of concrete, 197–200
Floors, 178–80, 188, 189, 191, 212–27
Flotation, 209–12
Formwork, 175, 177, 180, 234, 235, 240, 245, 278
Foundations, 206–8
Freeze–thaw, *see* Frost resistance of concrete
Frost resistance of concrete, 107, 154, 155, 167–9, 227, 272

Galvanized reinforcement, 104, 130, 154, 267–9
Gamma radiography, 303
Gap-graded aggregates, 159
Gradients (for roofs), 247
Ground water, 138, 141, 147, 204–12
Grouting of ducts, 269
Gunite, 156, 244, 260–6, 271, 273–81, 314, 315
Gunmetal, 131

Handbook to the Code, 6, 9, 10, 15, 30, 34, 35, 36, 37, 58, 63, 64, 67, 76, 81, 83–5, 89, 315, 316
Hardcore, 214, 215
Hardness of water, 148–51
Heat of hydration of cement, 170, 175–8
Heated concrete, 168
High alumina cement, *see under* Cement
High velocity water
  effect on concrete, 154–6
  jets for cutting concrete, 197, 198, 279
Hot weather concreting, 169–71
Humic acids, 147, 148
Hypalon, 121

Impermeability of concrete, 132–6

Joint fillers, 111
Joint sealants, 112–15, 187, 192, 252, 253, 292, 315
Joints in concrete
  construction (and daywork), 181, 188, 195, 196
  contraction, 178, 181–3, 222
  full movement, 187–90, 245, 251
  general, 187, 188, 222, 252, 279, 280, 289, 292, 294
  monolithic, 185, 190, 195, 196, 232, 253, 279
  partial movement and partial contraction, 188–93
  pin, 187, 188, 196
  sliding, 69, 178, 183, 184, 187, 188, 193–5, 232, 269, 296
  stress-relief, 178, 181–3, 193, 247

Langelier index, 149–51
Lean concrete, 214, 215
Lightweight aggregates, 103
Lignosulphonates, 107, 108
Lime, use in gunite, 77, 275, 276
Limestone aggregates, 103, 150, 151, 246
Linings, *see* Membranes
Loading tests, 307, 308
Lubricating oils, effect on concrete, 144

Mastic asphalt, 253, 254
Membranes, 243, 253, 254, 272, 290, 291, 295, 296
Metal roofs, 244
Metals, 127–31
Mill scale, 129, 130
Mix design, 156–60
Mix proportions, 219, 235
Modulus of elasticity, 336

Neoprene, 113

Non-destructive testing, 302–5
Non-ferrous metals, 130, 131
Notation, 3

Oil tanks, *see under* Petroleum oil
Organic acids, 147, 148
Organic impurities in aggregates, 101
Organic polymers, 117–21; *see also* Polymer resins
Oxy-acetylene flame for texturing concrete, *see* Flame texturing of concrete

Partial movement joints, *see under* Joints in concrete
Passivation of steel, 127, 128
Permeability of concrete, *see* Impermeability of concrete
Petroleum oil, 144, 294–6
  tanks, 294–6
PFA, *see* Pulverized fuel ash
pH, 128, 138, 139, 146–51, 205
Phosphor-bronze, 131
Piles, 206–10
Pipes and pipework, 228–31
Plain (unreinforced) concrete for walls, 291–4
Plastic cracking, 169, 171–4, 232
Plasticizers, *see* Admixtures for concrete
Pneumatically applied mortar, *see* Gunite
Polyester resins, 120, 140
Polyethylene sheeting, 178, 243
Polyisobutylene (PIB), 121, 253, 290, 291
Polymer concrete, 93, 121, 122
Polymer resins, 117–21, 253, 254, 265
Polypropylene, 93
Polysulphides, 113, 192
Polyurethane resins, 118, 119, 140, 192, 253, 254, 265, 290, 291
Polyvinyl acetate (PVA), 118, 121
Polyvinyl chloride (PVC), 117, 140, 244, 253, 254, 290, 291
Popouts, 227, 228
Portland cement, *see under* Cement
Post-tensioned concrete, *see* Prestressed concrete
Power floats and trowels, 225
Precast concrete, 248, 249, 271
Preload system, 260, 267, 270
Pressure-relief valves, 209
Prestressed concrete, 68–89, 248, 249, 259–73, 290
Protection against corrosion
  concrete, of, 131–52
  metals, of, 127–31, 261–9
PTFE, 183
Puddle flanges, 229–31, 235
Pulverized fuel ash (PFA), 110, 111, 234, 235
Pumping concrete, 286
Pundit, *see* UPV
Pyrites in aggregates, 287

Quality control and testing of concrete, 298–310

Rafts, 206
RAM (Rapid analysis machine), 301
Rebound, *see* Gunite
Reinforcement steel, 104, 179, 186, 220, 232–4, 247, 251, 269, 277, 292
Research organisations, 338–48
Retarders, *see* Admixtures for concrete
Roof slabs, 185–7, 193, 243–57, 269, 271–3
Roofing felt, 253, 254
Roofs, metal, *see* Metal roofs
Rust on reinforcement, 129, 233, 234

Salt content of aggregates, 102, 103
Schmidt hammer, 302, 303
Sea-dredged aggregates, 101
Sea shells in concrete, 102
Sea water, 151–3
Sewage, effect on concrete, 145
Shear connectors, 249, 250
Shotcrete, *see* Gunite

Shrinkage
  aggregates, of, 101
  concrete, of, *see* Drying shrinkage
Silicone rubbers, 113
Site concrete, 215, 216
Slip layer (sliding layer), 216
Sloping floors, 231, 232
Sludge tanks and sludge digestion tanks, 68, 143, 243, 244, 271
Slump of concrete, 158, 219, 231, 237, 247, 250
Sodium carbonate, 107
Sodium chloride, 105, 152, 153
Soft acid water, effect on concrete, 146–51
Spa water, 152
Specific gravity of cement and aggregates, 336
Specific heat, 337
Sprayed concrete, *see* Gunite
Stainless steel, 105, 130
Standards and Codes of Practice, 317–34
Steel, mild and high tensile, protection of, 127–30, 244
Stresses in concrete and steel, *see* Design
Styrene butadiene, 118, 119, 120, 154, 192, 235, 257, 265, 276
Sub-base, 213, 215, 216
Sub-grade, 213, 214
Sub-soil, 138, 204, 205
Sulphates, 141, 144, 151–4
Sulphur, effect on concrete, 145
Swimming pools, 47–67, 178
Synthetic rubbers, 121, 253

Tanks
  design in prestressed concrete, 68–89
  design in reinforced concrete, 18–46
  to hold petroleum products, 294–6
Temperature rise in concrete, 176–8
Temperature stresses, 71–4
Tensile strength
  concrete, of, *see* Design
  steel, of, *see* Design
Testing concrete, *see* Quality control and testing of concrete
Testing for watertightness, *see* Water test
Thermal baths, 152
Thermal contraction cracking, 169, 175–87
Thermal insulation of roofs, 254, 255, 271
Thermoplastics, 114, 192
Thermosetting compounds, 114–15
Tie bars, 179
Tolerances in construction, 349, 350
Toppings to concrete slabs, 249–53

Underdrainage, 208–10; *see also* Sub-soil
Unreinforced concrete for walls, 291–4
UPV (ultra-sonic pulse velocity) for testing concrete in structures, 304–5

Walls, 180–5, 189, 190–2, 232–43, 269–71
Water bars, 115–17, 181–96, 222, 237, 294
Water/cement ratio, 134, 140, 142, 150, 151, 153, 158, 159, 160, 206, 207, 219, 237, 246, 247, 276, 280
Waterproofing of roofs, 253–7
Water scars (water runs) in concrete, 238–40
Water test, 253, 257, 295, 308–10, 315
Watertightness, 209, 267, 281, 286, 315
Water towers, 281–91
Water treatment chemicals, effect on concrete, 146
Workability aids, *see* Admixtures of concrete

Yield of concrete mixes, 160, 161
Yield of mortar mixes, 161, 162

Zinc, 129, 130